Maxwell on Saturn's Rings

Maxwell on Saturn's Rings

Edited by
Stephen G. Brush,
C. W. F. Everitt,
and
Elizabeth Garber

The MIT Press
Cambridge, Massachusetts
London, England

Publication of this volume was supported by a grant from the General Research Board of the University of Maryland.

This book was set in Monophoto Times Roman by Asco Trade Typesetting Limited, Hong Kong, and printed and bound by The Murray Printing Company in the United States of America.

Library of Congress Cataloging in Publication Data

Maxwell, James Clerk, 1831–1879.
   Maxwell on Saturn's rings.

   Includes index.
   1. Saturn (Planet)—Ring system. 2. Astronomy—History—Sources.
I. Brush, Stephen G. II. Everitt, C. W. F. III. Garber, Elizabeth. IV. Title.
QB405.M38 1983   523.4′6   82-24890
ISBN 0-262-13190-0

Contents

Preface

Publication of the notes and letters of James Clerk Maxwell (1831–1879) has an intrinsic interest and value for historians of science, but the present set of documents offers something of wider interest. Maxwell's work on Saturn's rings is a case study in the production of first-rate mathematical physics for reasons that lie entirely outside the problem's intrinsic scientific merit. Maxwell had shown no previous curiosity about astronomical problems as an area of research while either a student at Cambridge or a Fellow of Trinity College. Only when the problem of the stability of Saturn's rings was presented as the subject of the 1856 Adams Prize Essay did Maxwell turn his attention to the question.

By 1856 Maxwell had already contributed to several areas of the physical sciences. By the age of twenty-five he had published a series of papers on color theory,[1] some papers on elasticity and the deformation of surfaces,[2] and presented two long and brilliant papers to the Cambridge Philosophical Society on electromagnetic theory in which he made his first attempt to give mathematical form to Faraday's conception of the electric and magnetic fields.[3] These last papers were later transformed into the work for which Maxwell is still remembered, but at the time his work on electromagnetism merely baffled most of his contemporaries. Maxwell appeared to have taken an already difficult domain of mathematical physics and made it more opaque rather than more transparent.

In short, Maxwell was a brilliant young man who had not as yet realized his potential in any one universally recognized piece of work. His electromagnetic theorizing seems to have created the impression that he was too eager to speculate without solid evidence. In contrast, his essay on Saturn's rings demonstrated a mastery of mathematical technique and an ability to analyze a problem into its elements and to prove the theorems essential to supporting the physical conclusions that led naturally from one element to the next. It clinched his reputation as a first-rate mathematical physicist.

This essay also gives us some idea of the mathematical methods actually used and the depth of their exploitation by a Cambridge wrangler in the middle of the nineteenth century, in contrast to those being explored by mathematicians and applied by engineers. So this volume is, in addition, a case study on the kind of work judged by nineteenth-century mathematicians and physicists as first-rate.

Maxwell's insight into the questions that needed to be addressed, his imaginative use of the available mathematical language, and his understanding of the limitations of his solution are indeed impressive. He seized on the central issue and brought to bear on it just the right amount of mathematical rigor to produce a conceptually simple solution. We can see just how complicated this solution became as we watch Maxwell spending many months working his way through the required mathematical details. His essay cost him much labor, as can be seen from the correspondence. His work was fit in around a heavy schedule of teaching, nursing his dying father, and then his marriage to Katherine Mary Dewar.

Maxwell's essay in celestial mechanics is a combination of deft handling of mathematical methods, conceptualizing the physics, and visualizing in real physical terms the results of the mathematics, demonstrating his understanding of how far the mathematics needed to be taken. This was a combination guaranteed to warm the heart of any academic critic and to impress the professional physicist or astronomer in any century. In the letters, especially those to William Thomson (later Lord Kelvin), we can see the ways in which Maxwell attacked the problem, including some of his false starts and errors, all too familiar in solving theoretical problems.

Astronomers and physicists may be interested in Maxwell's work on Saturn for another reason. All the sciences, especially the mathematical sciences, develop through the consideration of particular problems, not general issues, that then suggest the next generation of problems. Therefore scientists live in a world where immediate issues are of paramount importance and solved problems become, almost immediately, uninteresting. Their results are quickly integrated into the common knowledge of the discipline by being taught to graduate, and then undergraduate, students. In such a future-minded enterprise the complexities of the paths and their various twistings and dangers are often forgotten. Similarly, the origins of our presuppositions and assumptions are also buried along with the painstaking way in which they were built up through generations of observations, theoretical speculations, and mathematical confirmations. The problem of the nature and stability of Saturn's rings is a good case in point. Twentieth-century planetary astronomers—and so popularizers and media commentators on the Voyager rocket missions—can assume without question the results Maxwell so assiduously established. The rings are concentric and consist of small, solid satellites revolving about Saturn under the force of gravity. This case study is a reminder of the lengthy observational history and theoretical analysis that lies behind those results.

The Voyager missions relayed back to earth completely unexpected images of the whole solar system. Saturn is no longer a remote, still, marble-white, otherworldly sphere seen at a distance. Its rings are dynamic and complex, and the planet is richly textured. Simply put, the new technology of these missions has domesticated the solar system, making the planets appear far more like the earth. They all bear marks of a turbulent history, and in some cases carry visible signs of a violent present. These new, vital pictures have already erased memories of totally foreign, ordered, stable, and perfect images of earlier telescopic pictures of Saturn and its rings. To remind ourselves of the importance of available images in shaping the questions that theoreticians are even able to ask, we include some pictures that were within Maxwell's reach as he began work on Saturn's rings as well as some that became known to him only after he completed his essay. This gives us an idea of how Saturn appeared to astronomers, and hence the ways in which theoretical issues could be defined.

The core of this volume consists of Maxwell's letters to friends and colleagues as he worked on the essay and revised it for publication (documents 1–16); its final published version (document 17); and George Biddle Airy's review of it in the *Monthly Notices of the Royal Astronomical Society* (document 18). This is followed by Maxwell's correspondence with the Harvard astronomer George Bond (document 19) and his attempts (documents 20–28) to extend the analysis to the case where the particles making up the rings collide with one another as well as rotate about the planet, for which he used the methods of the kinetic theory of gases he had first published in January and June 1860.[4] (In his essay Maxwell had stated explicitly that this case was beyond the capabilities of the methods available in 1859.) The documents read like fragments from a manuscript never completed, marked by breaks in the logic of the argument from one to the next. Since there is no way of knowing at these points how Maxwell would have proceeded, the documents are separated at these breaks.

We introduce these documents with a historical sketch of observations of the rings and attempts prior to Maxwell to understand them. This is followed by an analysis of the circumstances that led to renewed interest in the problem in the 1840s. To give the immediate context of Maxwell's essay, we include here the correspondence among the Adams Prize Essay examiners in which they chose the topic and refined the wording of the published version. A study of Maxwell's published essay follows, emphasizing the originality of his exposition. Finally, we examine contemporary reactions to Maxwell's essay as a solution to the problem of Saturn's rings.

Much of this introductory correspondence might seem irrelevant to a collection of unpublished scientific papers; but while the science in it is dilute, it does reveal aspects of Maxwell not seen in his later letters. Maxwell's early correspondents were college friends rather than professional colleagues. The only ones who fit into the latter category in this early period of Maxwell's career are William Thomson, George Gabriel Stokes, and Peter Guthrie Tait. Tait, however, and Maxwell were old friends from their days as students at the equivalent of high school and then at Edinburgh and Cambridge universities. We have cut these letters when, in our viewpoint, they become irrelevant.

In the documents 〈 〉 indicates a word that was deleted in the original, ? a scribble that is indecipherable to all three editors, and [ ] our own insertions (usually a suggested title for some unpublished notes). Editorial notes to a document, indicated by superscript a, b, c, . . . , appear at the end of the document. These notes identify the correspondent and his relation to Maxwell, contemporaries of historical interest, and references to any of Maxwell's own published papers or works in progress.

We would like to thank Brigadier Wedderburn-Maxwell, N. H. Robinson of the Royal Society of London, and A. E. B. Owen of Cambridge University for giving us permission to publish this material and for assistance in obtaining copies of it.

Notes

1. "On the Theory of Colours in relation to Colour-Blindness," *Trans., Sect. Soc. Arts, Edin.* 4(1856), 394–400; "Experiments on Colour as perceived by the Eye . . . ," *Trans. Roy. Soc. Edin.* 21(1857), 275–298; "On the Theory of Compound Colour . . . ," *Phil. Trans.* (1860), 57–84; "Account of Experiments on the Perception of Colour," *Phil. Mag.* 14(1857), 40–47. These papers are reprinted in *The Scientific Papers of James Clerk Maxwell*, ed. W. D. Niven, 2 vols. bound as 1 (New York: Dover, reprint 1965), vol. 1, pp. 119–125, 126–154, 243–245, and 263–270, respectively.

2. "On the Equilibrium of Elastic Solids," *Trans. Roy. Soc. Edin.* 20(1853), 87–120; "On the Transformation of Surfaces by Bending," *Trans. Camb. Phil. Soc.* 9(1856), 445–470. The papers are reprinted in *Scientific Papers*, vol. 1, pp. 30–73 and 80–114, respectively.

3. "On Faraday's Lines of Force," *Trans. Camb. Phil. Soc.* 10[1856](1864), 27–83, reprinted in *Scientific Papers*, vol. 1, pp. 155–229.

4. "Illustrations of the Dynamical Theory of Gases," *Phil. Mag.*: part I, 19(January 1860), 19–32; part II, 20(June 1860), 21–37. Both parts are reprinted in *Scientific Papers*, vol. 1, pp. 377–409.

Maxwell on Saturn's Rings

Introduction

Interest in and speculation about the planet Saturn and its ring system has continued since it was first observed by Galileo Galilei (1564–1642) in 1610. He announced his discovery in a letter to Johannes Kepler (1571–1630), SMAISMRMILMEPOETALEUMIBUNENUGTTAUIRAS, intended as an anagram for ALTISSIMUM PLANETAM TERGEMINUM OBSERVAVI (I have observed the most distant of planets to have a triple form). The anagram was a common method among scientists in the seventeenth and eighteenth centuries of establishing priority without directly leaking any critical information, thus retaining a lead, if not a monopoly, in a particular area of research. Kepler misread the message as SALUE UMBISTINEUM GEMINATUM MARTIA PROLES (Hail, twin companionship, children of Mars), and this error seems to have been one source of the belief that Mars has two satellites, given literary form by Jonathan Swift in *Gulliver's Travels* (1726) and finally substantiated by Asaph Hall in 1877.[1]

Considerable controversy ensued over the precise nature of the appendages seen by Galileo. Only after using a considerably improved telescope was Christian Huygens (1629–1695) able to establish in 1656 that they were parts of a ring that surrounded Saturn whose diameter was rather more than twice that of Saturn. He also announced his discovery in 1659 as an anagram: AAAAAA CCCCC D EEEEE G H IIIIIII LLLL MM NNNNNNNN OOOO PP Q RR S TTTTT UUUUU, meaning ANNULO CINGITUR, TENUI, PLANO, NUSQUAM COHAERENTE, AD ECLIPTICAM INCLINATO (He is surrounded by a thin flat ring, which does not touch him anywhere and is inclined to the ecliptic). His observations were soon accepted despite some initial disagreement.[2]

Huygens's speculation, that the ring is a single solid structure that revolved with the planet while remaining at a constant distance from it, became more controversial after Jean Dominique Cassini (1625–1712) observed in 1675 a dark line separating the ring into two separate concentric ones.[3] A few years later Edmund Halley (1656–1742) concluded that Saturn rotated in the plane of the rings, a view confirmed by Sir William Herschel (1738–1822) in the 1780s from observations of the dark spots on the surface of the planet. From these observations he deduced that the period of rotation was almost $10\frac{1}{2}$ hours, while observations of certain bright spots on the rings showed they rotated with almost the same period. The problem of the number of rings, their divisions, and their separations remained a subject for observation well into the nineteenth century.

As observations continued and more details of the ring structure became known, theories of their nature multiplied. Jacques Cassini (1677–1756) first suggested that an infinite number of small satellites made up the rings, and this opinion was shared, without demonstration, by his father J. D. Cassini, Christopher Wren (1632–1723), Thomas Wright (1711–1786), and other astronomers at the beginning of the eighteenth century.[4] One of the least known among British and French astronomers, yet one of the more detailed, of these speculative theories was developed by Immanuel Kant (1724–1804). He suggested that the rings were formed by vapors ascending from the planet and are now composed of "particles" of unspecified size. Kant recognized that particles revolving in Kepler orbits around Saturn would move at speeds depending on their distances from the primary and that collisions between particles in streams moving at different speeds would tend to accelerate the slower ones while slowing the fast ones. This drag effect would break up the rings because the particles retreat from the centers of the rings. To avoid this instability Kant argued that the particles are segregated into noninteracting concentric circular bands.[5]

Pierre Simon, Marquis de Laplace (1749–1827), in 1787 was the first to bring the methods of the calculus to bear successfully on the problem of the structure of Saturn's rings and to establish the conditions for stability for at least a solid ring.[6] He argued that a stationary, solid ring would collapse under the gravitational attraction of the planet. He then considered a ring that rotated at such a speed that its centrifugal force balanced the attraction of the planet. Laplace also demonstrated the necessity for a system of concentric rings; a single solid ring would tend to fly apart. Gravitational theory alone was sufficient to establish this result independently of observations. But the theory also required that these rings be irregular and that their centers of gravity not coincide with their geometrical centers.[7]

Laplace later incorporated the rings in his suggestive theory of the origin of the solar system. He proposed that the planets were formed by the condensation of rings of vapor spun off from the extended atmosphere of a cooling, shrinking, rotating primeval sun. Similarly satellites were spun off from planets.

If all the particles of a ring of vapours continued to condense without separating, they would at length constitute a solid or liquid ring. But the regularity which this formation requires in all the parts of the ring, and in their cooling, ought to make this phenomenon very rare. Thus the solar system presents but one example of it; that of the rings of Saturn. Almost always each ring of vapours ought to be divided into several masses, which, being moved with velocities which differ little from each other, should continue to revolve at the same distance about the Sun.

After a planet has formed by the mutual attraction of the particles of vapor in a ring,

the planet resembles the Sun in the nebulous state, in which we have first supposed it to be; the cooling should therefore produce at the different limits of its atmosphere, phenomena similar to those which have been described, namely, rings and satellites circulating about its centre ... The regular distribution of the mass of rings of Saturn about its centre and in the plane of its equator, results naturally from this hypothesis, and, without it, is inexplicable. Those rings appear to me to be existing proofs of the primitive extension of the atmosphere of Saturn, and of its successive condensations.[8]

Something like Saturn's rings would be expected in the condensation process he proposed in the formation of the planets from the nebulous beginnings of the solar system. Reviewing the history of the subject in 1825, Laplace noted that according to William Herschel's observations, the ring rotates in 0.438 days while the planet itself turns in 0.427 days; he considered the faster rotation of the planet to be a confirmation of his cosmogonic hypothesis.[9]

During the first part of the nineteenth century, Laplace's theory was generally accepted, and it inspired some observers to report faint additional subdivisions of the rings that could not be confirmed later.[10] But a few theorists were skeptical and maintained that the rings were composed of either a coherent solid[11] or perhaps separate small particles.[12]

In 1849 Edouard Roche (1820–1883) published the first of a series of major works on planetary physics, including a discussion of the stability of a satellite under the tidal action of its primary. If the planet and satellites have equal density, the satellite cannot come closer than 2.44 times the radius of the planet without being broken up, and he noted that this distance is "a little farther than the external radius of the ring of Saturn" ("à peine supérieure au rayon extérieur de l'anneau de Saturn").[13] Unfortunately this result, later known as the "Roche limit," was published only in the *Mémoires* of the Montpellier Academy and was not generally known until after Maxwell completed his work on Saturn's rings.

During the years 1848–1849 there was a revival of interest in the rings in particular, among William Cranch Bond (1789–1859) and his son George Phillips Bond (1825–1865) at Harvard and William Dawes in England. Intensive work culminated in the nearly simultaneous discovery of an inner "dusky" ring (later called the "crepe ring") by the Bonds and Dawes.[14] It was noted in 1852–1853 by W. S. Jacob in India, William Lassell in England, and G. Bond and C. W. Tuttle that this ring is translucent, the limb of the planet being visible through it. This discovery, according to Alexander, "convinced astronomers that the whole question of ring structure needed a thorough re-examination."[15]

On April 15, 1851, G. P. Bond presented a paper at the meeting of the American Academy of Arts and Sciences in which he reviewed both observational and theoretical work on Saturn's rings. He concluded that a system of irregular solid rings such as Laplace had proposed would "become the source of mutual disturbances, which must end in their destruction, by causing them to fall upon each other." On the other hand,

the hypothesis that the whole ring is in a fluid state, or at least does not cohere strongly, presents fewer difficulties.

There being no longer an unyielding coherence between the particles of the inner and outer edges, they have not necessarily the same period of rotation about *Saturn*. A continual flow of the inner particles past the outer may be supposed, by which the centrifugal force will be brought into equilibrium with the other forces. And even should an accumulation of disturbances, of which the absence of inequalities lessens the probability, bring the rings together, the velocities at the point of contact will be very nearly equal, and the two will coalesce without disastrous consequences.

... Finally, a fluid ring, symmetrical in its dimensions, is not of necessity in a state of unstable equilibrium with reference either to Saturn or to the other rings.[16]

Benjamin Peirce (1809–1880) in remarks made following Bond's paper at the American Academy meeting and elaborated in a paper read to the American Association for the Advancement of Science in 1851 claimed to be able to show theoretically that the ring cannot be solid. Instead it "consists of a stream, or of streams of a fluid rather denser than water, flowing around the primary."[17] However, such a stream of fluid could not be maintained in stable equilibrium by the action of the planet alone, and it is stabilized by the perturbations of Saturn's satellites. Peirce did not present any detailed calculations to substantiate his claim, but ended the paper by confessing that his researches had just about caused him to abandon his former opposition to Laplace's nebular hypothesis.

On reading an abstract of Peirce's paper, Daniel Kirkwood was moved to publish his own remarks on Saturn's ring, originally presented in a lecture in Reading, Pennsylvania, on January 3, 1851. As a supporter of the nebular hypothesis, he saw the ring as representing "the most recent cosmical formation within the limits of the solar system" and suggested that it might in the future "collect about a nucleus and constitute a satellite."

The evidence of its solidity is not, I think, by any means conclusive. On the other hand, observations made within the last few years give a degree of plausibility to the presumption that it may be in a state of fluidity. I refer to the occasional appearance of dark lines, chiefly on the outer ring, which have been supposed to indicate a subdivision into several concentric annuli. They do not, however, appear to be permanent.[18]

The fact that faint subdivisions were seen to appear and disappear was taken as evidence by 1851 that Saturn's ring could not be a single solid body, but even more striking evidence was soon brought forward. In 1853 Otto Struve (1819–1905) reviewed measurements of the dimensions of the ring going back to Huygens, and concluded that "the inner edge of the interior bright ring is gradually approaching the body of the planet, while at the same time the total breadth of the two bright rings is constantly increasing."[19]

Throughout the decade Struve's conclusion remained controversial, accepted by some astronomers,[20] rejected by others,[21] and is now considered incorrect. Such debate renewed interest in Saturn's rings, whence it became a topic for the Adams Prize Essay at Cambridge University for 1856.

When Maxwell began to work on the stability problem, the ring system, as observed, considered of a bright outer ring (divided into at least two thinner rings), a bright inner ring, and within the latter the recently discovered dark ring. The planet, 74,937 miles in diameter, was surrounded by rings 100 miles thick whose outside diameter was some 170,000 miles. The mass of this ring system was believed to be about 1 percent of the planet's, and, there were indications of gradual changes in the ring structure. Theoretically the question of the nature of the rings was still open. They might still be a solid or liquid system or a collection of small bodies very close together. Only some of the conditions for the stability of a solid or liquid ring had been established.

The Adams Prize Essay had been instituted at Cambridge in 1848 to honor the astronomer John Couch Adams (1819–1892) for predicting the existence of the planet Neptune. It was awarded every two years for the best essay on some subject in pure mathematics, astronomy, or other branch of natural philosophy. The sort of problem selected for an essay subject required sophisticated mathematics and strong physical intuition, but not undue speculation. The question for the 1856 prize was the subject of an extended correspondence between two of the examiners, James Challis (1803–1882), Plumian Professor of Astronomy and Experimental Philosophy at Cambridge, and William Thomson (1822–1907), later Lord Kelvin, already professor of natural philosophy at Glasgow University. From this correspondence we learn the other subjects suggested and the kind of emphasis they expected in the analysis of the problem.[22]

The correspondence opens on February 28, 1855, with James Challis writing to William Thomson.

Cambridge University
Feb. 28, 1855
My dear Professor,

I am glad that you have consented to be an examiner for the Adams Prize. The Master of St. Peter's College desired me to inform you of your appointment by Grace of the Senate passed this day. The other examiner is Parkinson[23] of St. John's. As we have to give out a subject before the end of this term, I thought it better to lose no time in entering upon this part of our duty. The subjects hitherto proposed have been astronomical,—but they may be in "other branches of Natural Philosophy" & in "Pure Mathematics." We have not hitherto been successful in inducing competitors to come forward. On the first occasion there was but one, who gained the prize; on the next, there was also but one, & his Essay was too imperfect for a decision. On the last occasion there were no competitors; the subject was Biela's Comet. I fear that Cambridge mathematicians have no taste for investigations that require long mathematical calculations. I should be glad if you can suggest some subject that will be more likely to attract candidates. On the enclosed Paper I have put down ⟨three⟩ a few [?] The first of these was proposed in 1851, on which occasion the imperfect essay was sent in. Mr. Airy,[24] who was one of the Examiners, thought the subject a very good one, & recommended proposing it another time. I had rather this time make trial of a subject not requiring long astronomical calculations. The second subject, which also had Mr. Airy's approval, is I think too much like the one proposed in 1849, viz. the perturbations of Uranus & Neptune depending on the near commensurability of their mean motions. I think the third not a bad subject, if it can be definitely treated. It acquires an interest on account of the singular conclusions Otto Struve has recently come to respecting the approach of the inner Ring to the ball of Saturn. I have read his memoir, and consider the evidence to be satisfactory. If the subject be treated so as to decide whether or not the perturbations of extraneous bodies could give rise to such a change of form of the Rings, a step in science will be taken. The selection of subjects should have this end in view. It would also be interesting to ascertain the probable effect of the resistance of the ether on a body of the flat form of the Rings, as this enquiry would involve the consideration of the position of the flat surface with reference to the direction of the motion of the Rings *in space.*

I have put down subject (4) because I have never seen a satisfactory explanation of the aberration of Light.

I should be glad to have your remarks on the subjects I have mentioned, and to receive suggestions of other subjects that may occur to you.
Believe me,
My dear Professor,
Yours very truly
J. Challis
P. S. The Master of St. Peter's has probably informed you that all you have to do as Examiner may be transacted without the necessity of your being in Cambridge.

Accompanying the letter was the list of "Subjects suggested for the Adams Prize":

(1) An investigation of the perturbations of the moon in latitude produced by the action of Venus; and particularly of the secular movement, and the inequalities of long period in the movement, of the Moon's node.

(2) A complete investigation of the theory of the perturbations of two planets when their secular motions are accurately commensurate: especially in the case when the proportion of their secular motions is that of 2 to 1.

(3) An investigation of the perturbations of the forms of Saturn's Rings, supposing them to be fluid.
*** A comparison of modern with ancient measures of the external & internal diameters of the Rings of Saturn having made it probable that the forms of the Rings are undergoing change, it is required of the Candidates to ascertain, on the supposition of the fluidity of the Rings, what changes of form may be due to the perturbations of Saturn's satellites, and the perturbations of the Sun and Planets, and what would be the effect of a resisting medium, supposing the motion of the planet about the Sun to be combined with an assumed motion of the Solar System. The Rings may be supposed at a fixed epoch to be of very small thickness, circular and concentric with Saturn, & to be symmetrical with reference to the plane of Saturn's Equator.

(4) An explanation of the Aberration of Light.[25]

In his reply Thomson suggested three more possible subjects, the first involving the zodiacal light and perturbations of Mercury and Venus; the second, mechanical stability in general; the third, the elasticity of solids. Challis rejected these as being, respectively, too difficult to resolve without long numerical calculations, too general, and too far "removed from the general tenor of Canbridge mathematics" as well as being unconnected with astronomy. From all of this discussion the problem of Saturn's rings emerged as a compromise topic, acceptable to William Thomson and lying within the definition of subjects for the prize essay.

Cambridge Observatory
March 14, 1855

My dear Professor Thomson

I am much obliged by the attention you have given to the Adams Prize subject. Your letter contains most valuable suggestions and remarks which will be of use to me on future occasions. I proceed to give you my ideas on Nos. I, II, and III.

I like No. I, having some notions on the Zodiacal Light, which would be tested by the investigation of the perturbations of Mercury & Venus which you suggest. But it would require a very exact revision of the theories of these Planets to come to any result bearing on the Zodiacal Light, and long numerical calculations would be absolutely necessary, which, as I before stated, I am desirous on this occasion to avoid.

I think that some question under the head of Mechanical Stability (No. II) would be preferable, as requiring analytical rather than numerical calculation. But the general question of Mechanical Stability appears to me too large. Several of the special cases you have mentioned might of themselves be good Adams Prize subjects. For instance, "a careful examination of the demonstration given by Laplace of the stability of the Solar System," would be a useful and appropriate subject, that question being not yet set at rest. I am not certain that we should not do well in fixing on this for our subject, unless the one I shall presently speak of appears to have strong claims.

The Elasticity of solids (No. III) is a subject somewhat removed from the general tenor of Cambridge mathematics, and I fear we should get no answers, especially as experiments might be required, to which Cambridge men are not much given. I should like the subject on this occasion not to be wholly unconnected with Astronomy.

As you seemed to approve of the question about Saturn's Rings and to think that the discussion of it might lead to good results, I have drawn out a scheme, in conjunction with Mr. Parkinson, of instructions to the competitors, supposing that subject should be fixed upon. I should be glad to know what you think of it, and to receive any suggestion or alterations you may wish to make. In framing the instructions I thought it advisable to indicate several hypotheses respecting the constitution of the Ring, in order to exhaust the subject and allow of getting at some result positive or negative. It is probable that the hypothesis of rigidity is the only one admitting of exact treatment, and though we know beforehand that it is inconsistent with the condition of stability, the proof that it is so is not without interest, and may give scope for clever handling. I am of the opinion that the case of fluidity and of a surrounding atmosphere moving with the fluid, would not present insuperable difficulties, as in the simple form in which the Problem is proposed each particle both of the fluid and the air may be supposed (for a considerable interval at least) to be describing a circle uniformly about the axis of Saturn. This consideration virtually reduces the Problem to one in Hydrostatics. It seems to me that the treatment of the question must be very nearly the same whether the Rings be fluid and aeriform, or be supposed to consist of numerous small and unconnected masses.

I quite think that a definite result respecting the stability of the forms of the Rings may be arrived at by supposing no other forces to be concerned with the mutual action of the parts than that of gravity. As

soon as the friction of the parts, or a resisting medium enters into the
account, we have a cause operating to produce permanent change.
In the instructions to the candidates I have separated the part of the
Problem which may admit of a definite answer, from that which can
hardly be answered except upon gratuitous hypotheses. The latter part
may give rise to speculation and conjecture, which it may not be
useless to encourage.

I have difficulty in explaining how your "shower" should be so
opaque where it appears to cross the body of Saturn, and yet so dark.

The paper which contains the evidence for the gradual change of
forms of the Rings is one by Otto Struve published in the Memoires of
the St. Petersburg Academy VI Series Tom. V.

This term ends on March 30, and as the Paper containing the
announcement of the subject will have to be sent for your signature
before it is printed, you will see that there is not much time to spare.

Believe me,
My dear Sir,
Yours very truly
J. Challis

Prof. W. Thomson.[26]

In a final letter fixing the topic for the Adams Prize Essay, only changes
in the details of Thomson's suggestions remained to be hammered out.
Thomson not only accepted Challis's choice of topic but was becoming
intrigued by it himself.

Cambridge Observatory
March 23[d], 1855
My dear Professor Thomson,

I am glad that we agree about the Adams Prize subject. I have sent the
Paper as finally drawn up, which only waits for your signature. If you
will be so good as to return it by the first post, there will be sufficient
time for giving out the subject before the end of term.

You will perceive that we have adopted all your suggestions, with
the exception of omitting the word "varying" before "appearances."
It seems desirable to direct the Candidates to a consideration of the
actual appearances apart from any change to which they may be subject,
and in a particular manner to call their attention to the change which
appears to have the best support from observation, namely the change
of form.

The fact about the inner dark Ring is as you state. It was long since
noticed (by myself among others) as a dark stripe across the body of the
planet, not to be accounted for by the supposition of a shadow. The
recent observations have made out that this dark stripe is part of a *Ring*,
and that it is partially transparent, the contour of the Planet being dimly
seen through it.

I admit the force of what you say about the opacity of a shower of
opaque bodies. I have sometimes been astonished at the darkness induced by a snow shower. Taking into account that it does not require a
great degree of density of the component parts of the inner Ring to

render it opaque, we need not perhaps be surprised that it sends to us so little light.
Believe me, my dear Sir,
Yours very truly
J. Challis

Professor W. Thomson.[27]

In the published announcement of the competition (March 23, 1855), Struve's results were given much less prominence than Challis had first suggested. Instead of explaining why the rings *change*, candidates were expected to show under what assumptions they could be proved *stable*.[28]

James Clerk Maxwell (1831–1879) was descended from prominent Scottish families on both sides, and he inherited a large estate (Glenlair) near Dalbeattie, Galloway, in southwest Scotland, as well as the intellectual traditions of cultivated Edinburgh society. His father had trained as an advocate, but his passions were those of an improving landlord: the cultivation of his land, the care of tenants, practical technical matters, and science. Up to the age of ten Maxwell was educated at home, at first by his mother, until she died when Maxwell was eight years old, and then by tutors. Maxwell's formal education began at Edinburgh Academy, where Maxwell formed two life-long friendships, with Lewis Campbell (1830–1908), later a Greek scholar and the biographer of Benjamin Jowett and Maxwell (whose observations of these two subject were perceptive given the restrictions of Victorian prose); and Peter Guthrie Tait (1831–1901), a hard-headed practical mathematical physicist who as professor of natural philosophy at Edinburgh University was to set pedagogical duties above his own inclinations for research. Maxwell's first three years at the academy were undistinguished; then suddenly he came alive intellectually at the age of fourteen and produced his first scientific papers, on the geometry of ovals.[29] These papers marked the beginnings of Maxwell's style in science. In an age of analysis and algebra in physics, Maxwell's ideas, predilections, and explanations are essentially geometrical.[30] Maxwell entered Edinburgh University in 1847, at the normal age of sixteen, and studied there for three years, reading voraciously, doing mathematics, studying philosophy under William Hamilton (1788–1856) and physics with James David Forbes (1809–1868) before transferring to Cambridge in 1850, one year later than Tait. At Cambridge he became a pupil of William Hopkins (1792–1866), a remarkable teacher who trained George Gabriel Stokes (1819–1903), William Thomson, and Arthur Cayley (1821–1895) as well as Tait and Maxwell for the mathematics tripos. At Cambridge Maxwell became a member of the Apostles Club and was drawn into the circle around F. D. Maurice (1805–1872) and into Christian Socialism, whence came his committment to teaching working men in Cambridge, Aberdeen, and, later, London.

In 1860 Marischal College was merged with Kings College, Aberdeen; as a result Maxwell lost his chair and moved to the chair at Kings College, London.[31] He resigned from this position in 1865 and lived privately, writing his great *Treatise on Electricity and Magnetism*, until he emerged from retirement in 1871 to become the first professor of experimental physics at Cambridge. He planned and then directed the Cavendish Laboratory until his death from abdominal cancer in November 1879.

Maxwell's early scientific work was in the theory of elastic solids. This excursion into continuum mechanics was significant for his researches into electromagnetic theory; the connection was already apparent in a paper published in 1856. Part I of this paper was an exposition of the analogy between lines of force and stream lines in an incompressible fluid.[32] He then extended the analogy to consider a resistive medium through which the fluid flows and the effect when the fluid flows across a boundary into a medium of differing porosity. This approach by analogy not only was helpful in calculating processes in magnetic materials and dielectrics but enabled Maxwell to distinguish magnetic induction from magnetic force. He concluded this first part by exploring the parallels between electric currents and magnetic lines of force. Electromagnetism proper was treated in part II, wherein he obtained a complete set of equations, in cartesian and not vector form, between the vectors for the electric and magnetic forces and the current density and magnetic induction.[33] He developed an integral representation of the field equation, the classification of vector functions into forces and fluxes, and a symmetry between the force vectors. Although his later papers on electromagnetism and his *Treatise* would overshadow this first effort, Maxwell returned to all these themes in his later work, even to using the analogy of fluid flow.

This bold synthetic work was largely incomprehensible to his contemporaries, who could not understand his insistence on mathematizing Michael Faraday's (1791–1867) sophisticated and essentially geometrical ideas on electric and magnetic forces.[34] Maxwell therefore had yet, at the age of twenty-six, to establish his reputation. His work on color theory was interesting, but not directed toward what was seen as a vital problem in physics; his work in electromagnetic theory might be useful or merely a flight of scientific fancy. A solid piece of mathematical physics solved effectively yet imaginatively was necessary to consolidate a rather shaky career. When Maxwell began work on the problem of Saturn's rings, he had already resigned his fellowship at Trinity College, Cambridge (1856), to take the chair in natural philosophy at Marischal College, Aberdeen.

The Adams Prize Essay was the perfect opportunity for demonstrating mathematical facility along with just the right amount of physical intuition.

It was also very difficult. Maxwell's entry was the only one in the prize competition. On December 31, 1856, Challis wrote to Thomson,

I received Dec. 17 from the Vice-Chancellor one exercise for the Adams Prize. It is well deserving of attention, and indeed from the consideration I have hitherto given to it, I think very favorably of it. I shall
soon have completed my examination of it, and shall be glad to know
whether I shall forward it to you as soon as I have done with it. Mr.
Parkinson is willing to take it when it suits your convenience. The
weight of the Ms being about 12 oz, I find that according to the postal
arrangements, it can be sent in an enclosure open at the ends at the
charge of 4^{d}.[35]

The Adams Prize was awarded to Maxwell in June 1857.[36] Soon afterward he had the opportunity to discuss the subject with Challis and Thomson, and the published version of his essay incorporates additional work done in 1857 and 1858 (see reprinted letters, documents 2, 3, and 7).

The results of Maxwell's investigation have been summarized in Airy's review (document 18) and in Campbell and Garnett's biography (1882), but neither source gives much indication of Maxwell's methods. The most striking feature of his essay is the simplicity of its mathematical techniques. It can be understood completely with a knowledge of linear differential equations, simple potential theory, and Taylor's theorem together with elementary Fourier analysis. His essay is divided into three main parts, together with an introduction and a conclusion. In part I he considers the stability of irregular solid rings; in part II, fluid rings and rings composed of independent satellites; and in part III, the stability of two rings of satellites and under mutual perturbations. Our discussion will be based on the version of the essay published in 1859.

Maxwell's sureness of touch is illustrated by his derivation of the equations of motion for a solid ring. He begins by defining the positions of the ring and planet by three variables (figure 1); r, the distance between the centers of gravity of the ring and the planet (at R and S, respectively); the angle formed at G between the line SR and a fixed direction in space; and the angle between SR and a fixed direction in the ring. At first sight there seems little to this, but actually the right choice of variables is important in obtaining a neat set of fundamental equations. The next step is to determine the gravitational forces between the ring and Saturn. Here Maxwell follows the usual approach of working with potentials rather than forces, but instead of calculating the gravitational potentials at different parts of the ring due to the planet, he inverts the problem and determines the potential at the planet due to the ring. This is a master-

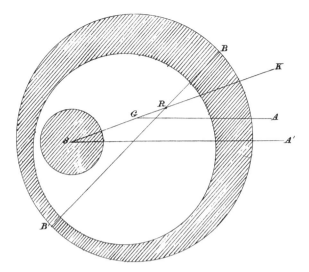

Figure 1
Establishing the variables for a solid ring. From Maxwell's essay for the Adams Prize.

stroke. For there is a standard theorem in gravitation that a spherical body can be treated as a point mass concentrated at its center. Consequently it is only necessary to calculate the potential at one point, the center of Saturn. This leads to a general expression for the forces and, using Newton's laws of motion, to the equations of motion for a ring of any form.

From the equations of motion, Maxwell derives two general conditions on the potential if the motion is to be uniform that immediately set some restrictions on the form of the ring. Thus Maxwell is able to examine whether small disturbances of the ring are without effect or grow without limit (as in the example considered by Laplace) and precipitate the ring onto the planet. Without detailing every step, it is sufficient to say that Maxwell expands the potential to first order by Taylor's theorem and derives a second set of equations that describes the motion of the ring under small displacements. These equations yield certain additional restrictions on the potential that must be satisfied for stable motion.

Maxwell then investigates the implications of these restrictions on the potential for the shape of the ring by means of Fourier's theorem. (He had read Fourier's *Théorie analytique de la chaleur* while an undergraduate at Edinburgh.) He does so by translating the restrictions on the potential into restrictions on the magnitudes of the first three coefficients in a Fourier analysis of the shape of the ring. To determine whether a partic-

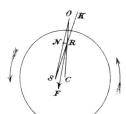

Figure 2
Stability of a loaded, solid ring. From Maxwell's essay for the Adams Prize.

ular ring is stable, he has simply to calculate its first three Fourier coefficients and to substitute them into the general formulas.

Maxwell finds several particular forms of a solid ring unstable except in one curious special case: a uniform ring loaded at a single point with a heavy satellite whose weight lies between 4.43 and 4.81 times the weight of the ring. In an intuitive argument to show why such a ring may be stable, Maxwell explains that the attraction of the planet acts at a point outside the ring (figure 2). This implies that there is a couple as well as a force acting on the ring that tends to alter the moment of inertia. If the load is adjusted correctly, the two effects balance one another and the motion is stable.

Since Maxwell's analysis is given in terms of Fourier coefficients, he proves quite generally that any solid ring must be extremely lopsided to be stable. However, the observed rings show no signs of lopsidedness, and in any case an extremely irregular ring would be certain to collapse under gravitational forces. Therefore the theory of a solid ring must be abandoned. But the motion of a solid ring is simple compared to one whose parts are movable. Consider a circle of satellites equal in size revolving around a planet and suppose that one of them accidentally becomes displaced in any direction. Because the satellites attract one another, the equilibrium of forces will be disturbed for not only the displaced satellite but also all the others. Consequently their motions will be disturbed, further altering the forces, and eventually the whole configuration will be deranged. Even if this case can be calculated, the actual problem is still far from solved because Saturn has not one ring but many, each of which attracts the others. Moreover it is very unlikely that the bodies forming any ring would be of equal size.

The investigation of the stability of such a system seems impossible, yet again Maxwell uses the right strategy. Instead of studying individual satellites, he applies Fourier analysis to investigate the propagation of waves in a ring of satellites and demonstrates that under certain conditions

Figure 3
Radial waves in a ring of 36 satellites: left, wave of fifth order; right, wave of eighteenth order.

a complete series of waves can exist without causing collisions between the satellites. Since any disturbance can be resolved into a system of waves, this means that small accidental disturbances will not disrupt the ring. Fourier's method contains two major advantages for analyzing the problem: It makes it relatively simple to calculate the forces between different parts of a ring containing a wave, and the number of undulations in a ring of satellites is limited. In the following account the number of undulations in the ring will be termed the order of the wave. (Figure 3 illustrates waves of the fifth and eighteenth orders in a ring of 36 satellites. The wave of eighteenth order, in which neighboring satellites oscillate in opposite directions, contains the maximum number of undulations for this ring.) The maximum number of undulations in any ring is equal to half the number of satellites.

Before proceeding further Maxwell establishes the variables for a single satellite in a uniform circular ring rotating with constant angular velocity (figure 4). *S* is Saturn, *SA* is a direction fixed in space, and *SB* is a radius rotating at the same speed as the ring, so that the angle *ASB* increases uniformly with time. *P* is the mean position of a particular satellite, so that the angle *BSP* defines the position of the satellite in the ring. This angle, which will be called *S*, is independent of time. Any displacement of the satellite from *P* can be resolved into the three components ρ, σ, ζ, which will be termed the radial, tangential, and normal displacements, respectively. The object of the calculations is to determine how ρ, σ, and ζ vary with time t and the angular position s. Since Maxwell is using Fourier's methods, the results will be expressed in terms of radial, tangential, and normal waves, corresponding to displacements in the three

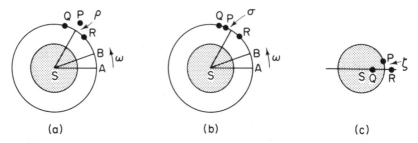

(a)   (b)   (c)

Figure 4
Parameters for describing the position of a single satellite displaced from its equilibrium position (ρ, σ, ζ).

directions. However, if we consider the small accidental displacement in the normal direction illustrated in figure 4c, the displaced satellite experiences a force in the direction opposite to the displacement because of the attractions of Q and R. In the tangential case (figure 4b) the attraction to R is decreased, so that the net force is in the same direction as the displacement. The attractions tend to counteract small normal displacements but enhance small tangential displacements, and the tangential displacements lead to instability in the ring's structure. However, Maxwell demonstrates that in a rotating ring the tangential motions become coupled with the radial ones in such a way as to counteract this instability.

Maxwell's next step is to calculate the forces between different parts of a disturbed ring. For the present problem, he finds it more convenient to work with forces rather than the potential functions of the solid-ring case. In a few simple steps he determines the three components of force on a given satellite due to radial, tangential, and normal disturbances of any wavelength. These expressions contain five complicated trigonometrical functions and six unknown constants, but they are completely general in form, and from these Maxwell sets up the equations of motion for displacements in any direction. The equations for normal waves confirm the conclusion already given, that normal displacements will will not lead to instability.

He has now reached the heart of the problem. Confining his attention to motions in the radial and tangential directions, Maxwell demonstrates that the two kinds of motion are coupled. He then obtains a pair of equations for the relative amplitudes of the radial and tangential waves and another equation for the frequencies associated with a given wavelength. The equation for the frequencies is a biquadratic; that is, it has four different solutions grouped in two pairs. If the mass of Saturn is sufficiently large, the four solutions are real and correspond to four

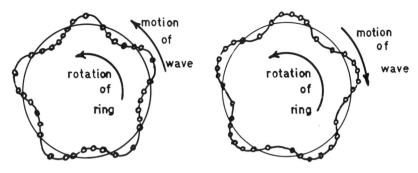

Figure 5
Waves of types 1 and 4 (left and right, respectively) in a ring of 36 satellities.

different types of waves. Waves of types 1 and 4 are almost identical
but travel around the ring in opposite directions (figure 5). For both
waves, each satellite describes an elliptical path about its mean position,
moving in the sense opposite to the direction of rotation of the ring. For
a wave of type 1 the satellites crowd together at the points nearest to
Saturn and separate upon moving outward; for a wave of type 4 the
greatest crowding occurs at points farthest from the planet. The velocities of the two waves relative to the ring are given approximately by
dividing the rate of rotation of the ring by the number of undulations.
Thus we may consider the wave of fifth order (figure 3) and take William
Herschel's figure of $10\frac{1}{2}$ hours for the period of Saturn's rings. Viewed
from a point within the ring, the two waves travel in opposite directions
and take $52\frac{1}{2}$ hours to move full circle. Viewed from a stationary point,
on the other hand, both waves travel forward, but type 1 will take 8.75
hours to complete the circle and type 4 will take 13.1 hours.

For the benefit of "sensible image worshippers" (document 10), Maxwell
had a model constructed to show the motions of waves of types 1 and 4
in a ring of 36 satellites (figure 6; illustrations of the model are given
with descriptions in documents 10, 16, and 17). A pair of waves of types
2 and 3 travel at less than 1 percent of the velocity of the pair of waves of
types 1 and 4. For waves of types 2 and 3 the satellites again describe
elliptical paths about their mean positions, but the ellipses are greatly
elongated. Viewed from a stationary point in space, each of these satellites
moves in an almost circular orbit about Saturn whose radius gradually
increases and diminishes during a period of many revolutions. The most
likely sources of disruption, then, in a disturbed ring are the tangential
motions of the waves of highest order, since in such waves neighboring
satellites are brought very close together. Maxwell did find that the condition for stability places a restriction on the total mass of the ring; for

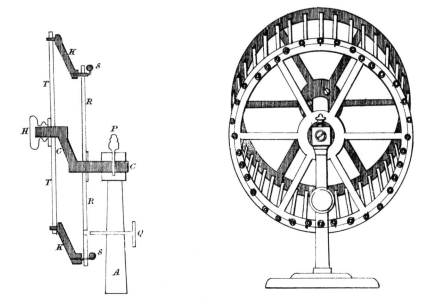

Figure 6
Model to illustrate waves of types 1 and 4. From Maxwell's essay for the Adams Prize.

example, the total mass of a ring of 100 satellites must not exceed 0.02% of the mass of Saturn.

Maxwell's discussion then shifts to the problem of why a rotating ring is stable despite the inherent instability of the tangential attractions. For each satellite in a ring, free from disturbances and rotating at uniform speed, the attraction of Saturn and the centrifugal force due to rotation will be exactly in balance. If one of the satellites is displaced forward, the centrifugal force will increase and the satellite will move outward until eventually it is traveling in an appreciably larger orbit. This satellite will fall behind the other members of its original ring even though its linear velocity is greater than theirs. After falling behind a certain distance, the attractions of the other satellites will cause a retardation, the centrifugal force will decrease, and the satellite will start moving inward again. Thus in a rotating ring the instability with respect to tangential displacements is overcome by coupling of the radial and tangential motions. Since the radial motions are controlled by the mass of Saturn and the tangential ones by the distribution of masses in the ring, they appear in Maxwell's stability condition. He also derives another form of the stability condition in which the magnitude of the tangential attractions is related to the rate of rotation of the ring. As expected, the condition is of such a form as to make a stationary ring always unstable.

Maxwell also determines the nature of the motions when the stability condition is not fulfilled. The satellites still trace elliptical paths about their mean positions, but the amplitudes gradually increase and the ellipses tilt forward until the satellites collide with one another and the ring is thrown into confusion.

The sizes of the satellites has not yet entered into the discussion; Maxwell's calculations so far have been concerned with the stability of a single ring containing both large and small satellites. He concludes that the large bodies become symmetrically distributed about the ring with strings of finer particles between them. The motions of the larger satellites are identical with those occurring in a uniform ring; but the waves in the small particles, while still stable, are split up and reflected at the discontinuities in the ring formed by the larger satellites.

We must now think more carefully about the actual astronomical observations of Saturn. Maxwell has demonstrated that a ring of satellites can be stable; but the observed rings are clearly not just single circles of particles, since each one is many hundreds of times broader than it is thick. Perhaps the observed rings are formed of many concentric rings; alternatively they may be continuous fluid sheets. In both cases there are two further possibilities: Either the inner and outer halves are bound together so strongly by attraction that the ring rotates as a whole or the different regions slide over one another with more or less the same speeds as isolated particles. In the latter case the inner edge of the ring would rotate more rapidly than the outer edge.

Rings in which the inner and outer parts rotate with the same angular velocity may be called semirigid. Laplace had proved that a truly rigid ring will fly apart unless its density exceeds 0.8 times that of the planet, as the centrifugal and gravitational forces cannot remain balanced throughout the ring. The same argument applies to semirigid rings. But such rings may also be destroyed if unstable waves are propagated within them. Maxwell's conclusion is that a semirigid ring of fine particles will break up under tangential forces unless its density is lower than 0.003 times that of the planet. Similar arguments give the maximum density for a continuous liquid ring as 0.024 times that of the planet. Since these conditions are incompatible with the one given by Laplace, both types of semirigid ring are unstable. Hence there seems to be only one possibility: concentric rings of satellites, each rotating with the speed appropriate to its own radius. But these concentric rings cannot be treated independently, for they attract one another. In the last section of the essay, Maxwell deals with the mutual perturbations of two rings.

He begins by noting that the additional forces on any ring caused by the presence of a second ring can be divided into three parts: the constant attraction occurring when there are no waves in either ring; the variation in the attraction on one ring due to its own disturbances, which alter its position relative to the other ring; and the variation in the attraction on one ring due to disturbances in the other. He discusses these in turn. The constant attraction tends to make the outer ring revolve faster and the inner ring more slowly than would otherwise be the case. The effect of the constant attraction of one ring on the disturbances in the other is to alter the velocities of the two fast waves (waves of types 1 and 4) and to leave the slow waves (types 2 and 3) unaffected. The velocity of the fast waves is reduced on the inner ring and increased on the outer. Finally, the third part of the additional force causes the four types of free waves in one ring to induce four types of forced waves in the other, so that eight different kinds of waves may be propagated in each ring. With more than two rings the number of different kinds of waves is proportionately increased.

Without all the technical details we can still appreciate the elegance of Maxwell's calculations. First he obtains the radial and tangential forces on one ring due to radial and tangential displacements of the other. These lead to equations for the amplitudes of the waves in the first ring, which are very similar to the corresponding equations already obtained for a single ring, except that they contain additional terms due to the motions in the second ring. If a direct method of solution were attempted, an equation of the eighth degree would be obtained yielding the velocities of the eight different kinds of waves. This equation would be analogous to the biquadratic equation for the frequencies in a single ring, but being of the eighth degree, it would be prohibitively difficult to solve. Here Maxwell notes that for physical reasons the additional terms are small. Consequently, using Maclaurin's theorem, he is able to obtain approximate solutions for the wave motions in a pair of rings. The motions are nearly always stable, but for certain ratios of the radii, a wave of one type in one ring may come into resonance with a wave of a different type in the other. The two waves strengthen one another and grow indefinitely until both rings are thrown into confusion. If the actual rings of Saturn are formed of concentric rings separated by very small distances, it is almost certain that some will be at the right distances for unstable motions. It appears that a few rings will break up, and this means that their particles will fly off in all directions and collide with the members of the other rings. We can go no further with the methods of analysis available, as Maxwell points out.

When we come to deal with collisions among bodies of unkown number, size, and shape, we can no longer trace the mathematical laws of their motion with any distinctness. All we can now do is to collect the results of our investigations and to make the best use we can of them in forming an opinion as to the constitution of the actual rings of Saturn which are still in existence and apparently in steady motion, whatever catastrophes may be indicated by the various theories we have attempted.[37]

In the ensuing discussion Maxwell highlights several factors that act to reduce the destructive interference between rings. The particles in any ring certainly vary in size, and this tends to impede the free propagation of the waves around the ring. It is also possible that the rings are grouped in narrow bands in such a way as to reduce the perturbations. Nevertheless some interference will probably take place and energy will be lost in collisions. Maxwell's final calculation yields the nature of the general changes in a system of rings in which the angular momentum remains constant while the total energy gradually diminishes. He concludes that there is a tendency for the outer rings to move outward and the inner to move inward.

When Maxwell started to work on this problem, astronomical opinion seemed to favor the theory of fluid rather than solid rings.[38] However, Maxwell apparently did not become aware of this shift in opinion, and in particular, of the papers of Bond and Peirce, until after he had completed most of his essay.[39] He devoted only a few pages to investigating the stability of a fluid ring, concluding that "a continuous liquid mass cannot revolve about a central body without being broken up, but that the parts of such a broken ring may, under certain conditions, form a permanent ring of satellites.[40] He returned to the fluid ring in a later section of the essay in studying possible causes of the long-term changes in form of the rings, and it was here that he quoted the data from Stokes on the viscosities of air and water. Though he alluded to Peirce's work in a last-minute footnote, he did not consider Peirce's suggestion that fluid rings could be stabilized by the action of Saturn's satellites, perhaps because no detailed justification of this suggestion had been published.[41] Document 2 indicates his skepticism about Peirce's theory.

Maxwell's achievement has been summarized as disproving a generally held belief that the rings are continuous solids[42] and proving that rings composed of many small solid particles may be stable.[43] It is true that Maxwell finally disposed of the solid-ring theory, but inaccurate to claim that this theory was generally accepted when his essay was published in 1859. It is also correct that Maxwell demonstrated the stability under various disturbing forces of a system of rings composed of particles; in particular, he showed that the tendency toward conglomeration into a single satellite, suggested by the nebular hypothesis, would be effectively counteracted by the dynamical factors involved in the revolution of the particles around the massive central body.[44] He also concluded that the changes in size of the rings determined by Struve were compatible with his theory, although it might also be necessary to invoke the action of a resisting medium.[45] The essay ends with the suggestion that even more radical changes in the appearance of the rings may eventually be observed:

If the changes already suspected should be confirmed by repeated observations with the same instruments, it will be worth while to investigate
more carefully whether Saturn's Rings are permanent or transitionary
elements of the Solar System, and whether in that part of the heavens
we see celestial immutability, or terrestrial corruption and generation,
and the old order giving place to new before our own eyes.[46]

Maxwell's results were presented to the Royal Astronomical Society in a long abstract prepared by Airy, the Astronomer Royal (document 18). In his outline Airy gave considerable attention to Benjamin Peirce's work, apparently to refute suggestions that Peirce had solved the problem before Maxwell. After summarizing Maxwell's investigations and quoting his conclusion that the rings must consist of "an indefinite number of uncon-nected particles," Airy claimed that "the theory of *Saturn's* rings is now placed on a footing totally different from any that it has occupied before, and that the essay which we have abstracted is one of the most remark-able contributions to mechanical astronomy that has appeared for many years."[47]

Despite such powerful support, Maxwell's work was not immediately recognized by astronomers. John Herschel ignored it in successive edi-tions of his authoritative text, *Outlines of Astronomy*, and Benjamin Peirce failed to mention it in communications to the American Academy of Arts and Sciences in 1861–1862.[48] When G. A. Hirn presented his own theory of the particulate constitution of Saturn's rings to the Paris Academy of Sciences in 1872, Airy quickly pointed out Maxwell's priority.[49] Hirn, however, defended himself against a possible charge of plagiarism by claiming that not only was he ignorant of Maxwell's work, but so were most astronomers:

With regard to my ignorance concerning M. Clerk-Maxwell's work, I
shall not try to justify myself; I shall say only that before the appearance
of my memoir, I had carefully read many recent works completely,
written by both eminent astronomers and popularizers who are up-to-
date with the most recent progress of science, and I was never able to
find a hint of the existence of the fine work of M. Clerk-Maxwell.
  This is easily explained by the difficulty that even the most active
scientists have today to keep themselves up-to-date with the very numerous publications in scientific literature, whatever their value.

(En ce qui concerne mon ignorance au sujet du travail de M. Clerk-Maxwell, je n'essayerai pas de m'en excuser; je dirai seulement qu'avant de faire paraître mon Mémoire j'avais soigneusement lu plusieurs ouvrages tout récents, écrits soit par des astronomes éminents, soit par des vulgarisateurs parfaitement au courant des progrès les plus récents de la science, et je n'y ai rien trouvé qui pût me faire soupçonner l'existence du beau travail de M. Clerk-Maxwell.

Cela s'explique aisément par la difficulté qu'ont aujourd'hui les savants les plus actifs de se tenir au courant des productions si nombreuses de la littérature scientifique, quelle que puisse être leur valeur.)[50]

On the other hand, in 1863 G. P. Bond corresponded with Maxwell on Saturn, so he probably knew of the essay. One of the first astronomers to comment on Maxwell's worth was another American, Daniel Vaughan (1818–1879). Vaughan had suggested earlier that tidal forces would disrupt a satellite in an orbit close to a large planet.

The rings of Saturn furnish an instance of matter revolving around a centre in a region so dangerous to satellites . . . . At the mean distance of the outer ring, a satellite as dense as Saturn would be incapable of retaining its planetary form; nor could one of double the density revolve in security in the space which the centre of the ring occupies . . . . There is also much reason to believe that the rings of Saturn are the remains of two former satellites.[51]

In a later paper Vaughan noted Maxwell's result that a solid ring could not be stable unless "loaded" with a bump containing about $4\frac{1}{2}$ times as much matter as the rest of the ring and his conclusion that fluid rings would break into satellites unless their density were less than 1/42 of that of the primary. Vaughan quarreled with this conclusion because his own earlier results indicated that at this distance a much more dense liquid ring was necessary in order to survive the disruptive tidal forces of the primary.

In investigating the case of a ring of numerous solid satellites, or fragments, Maxwell finds a combination of very extraordinary conditions necessary to prevent the derangements and permanent changes which collisions and friction are expected to occasion. The bodies are to be all equal in mass, and placed in regular array around Saturn; but the intervals between them must be very great compared with the linear dimensions; and the ratio between the planet and the ring must, according to his formulae, be greater than .4352 multiplied by the square of the number of satellites composing the latter. When we consider the vast number of such bodies required to maintain the continuity of the ring, and the great improbability that all the immense group should have the peculiar conditions for preventing one from striking another, we may regard the essay of the eminent mathematician as a proof that the disconnected matter composing the annular appendage, whether it be fluids or solids, cannot be maintained in its present condition without the occurrence of friction and collisions between its parts.[52]

Other early favorable notices of Maxwell's work came from T. W. Webb in a series of articles that gave a detailed review of the state of observational knowledge of Saturn.[53] He noted that Secchi's idea that the rings were dissipating had been shown to be improbable by Main and Kaiser. In discussing other theories of the rings, Webb noted that Maxwell had demonstrated that they could not be solid or fluid but were composed of unconnected particles; however, "to obtain ocular proof of the accuracy of this remarkable theory, to which so much publicity has recently been given by the ingenious and elaborate treatise of Mr. Proctor, may never perhaps be granted to us . . . . It remains to be seen what may be effected by persevering examinations; hitherto it must be admitted that instead of removing, it has increased the difficulties of the subject".[54] Webb concluded that the nature of the rings remained conjectural; the observations were both confusing and contradictory.

The ingenious and elaborate treatise to which Webb referred was published by the prolific British astronomer and science writer Richard Proctor, who, with Airy, was the strongest proponent of Maxwell's theory in the few decades after its publication. Proctor presented Maxwell's theory in *Saturn and His System*, first published privately in 1865, but widely circulated in a second edition of 1882. He also noted, later, that Maxwell had proved the rings to consist of "flights of minute bodies, each travelling on its own orbit," and added that some readers of his book seemed to have the impression that this theory was his own, even though he gave Maxwell full credit.[55] By 1880 other astronomers had accepted the view that Saturn's rings are composed of streams of solid particles and acknowledged Maxwell's role in providing a theoretical foundation for this view.[56]

The spectroscopic study in 1895 by the American astronomer James Keeler is now regarded as "the classic proof of James Clerk Maxwell's theoretical prediction that the rings of Saturn are meteoritic in nature."[57] Subsequent research has refined and elaborated Maxwell's investigation while confirming his main conclusions.[58]

While Maxwell published nothing else on Saturn's rings, he did work on the problem, specifically, the unfinished aspect of his analysis: the effects of collisions between satellites. Maxwell invented a method that held promise of applicability to this problem with his first kinetic theory paper, published in 1860.[59] In this paper he considered the collisions of an infinite number of moving small, hard, solid spherical particles, introduced probability arguments into the solution (their first appearance in the solution of a physical problem), obtained the Maxwell distribution for velocities, and from this deduced the mean free paths of the particles and the pressure they exerted on a unit area of the containing vessel. More important, Maxwell linked the mean free path and his distribution function to the measurable transport coefficients of gases, viscosity, diffusion, and thermal conductivity.

Thus for the first time physicists were able to measure the molecular characteristics of gases. Maxwell himself performed a series of simple yet elegant experiments to measure the viscosity of air over as wide a range of temperatures as was reasonably available to the physicist in the early 1860s. These experiments were the subject of the Bakerian Lecture at the Royal Society in 1866.[60]

During the same period Maxwell published two more important papers in electromagnetic theory,[61] extended his theory of colors, and, reexamined his paper on Saturn's rings. These notes (documents 20–28) remained as fragments, some more organized than others, and unpublished. He tried to consider the problem of the collisions of rough particles confined to a thin ring revolving about Saturn. That this approach was unsuccessful is not surprising, since Maxwell was attempting to generalize the problem of the collision of particles to the inelastic case while restricting the allowable motions of the particles themselves (namely, to a thin ring held in an orbit by gravitation). The collisions had to be inelastic because, as Maxwell pointed out, if they were elastic, the rings would disperse into a cloud. The ring shape was maintained by the gravitational field of Saturn, which kept the particles moving in their restricted paths. In order to obtain an integrable equation for the velocity distribution function, Maxwell had to eliminate the condition that the collisions be inelastic. In a second attempt Maxwell sought to advance his earlier results by changing variables. This approach got him no further along toward a solution, but it is indicative of his clear grasp of the limitations of his published analysis and what in principle had to be done to complete it. It has been suggested that it was the investigation of Saturn's rings that led Maxwell to the kinetic theory of gases. The documents do not establish such a direct connection. Indeed, the opposite may be inferred. Kinetic theory led him to reconsider aspects of the problem of Saturn's rings.

Soon after these notes were written Maxwell abandoned his first kinetic theory and developed a second, in which the particles acted as centers of force[62] and collisions were replaced with encounters. This vastly improved his derivation of the velocity distribution function. But as far as we know, he never returned to the problem of Saturn's rings.

Notes

1. Owen Gingerich, "The Satellites of Mars: Prediction and Discovery," *Journal for the History of Astronomy* 1(1970), 109–115; Albert Van Helden, "Saturn and His Anses," *Journal for the History of Astronomy* 5(1974), 105–121.

2. Albert Van Helden, "'Annulo cingitur': The Solution of the Problem of Saturn," *Journal for the History of Astronomy* 5(1974), 155–174; "Christopher Wren's *De Corpore Saturni*," *Notes and Records of the Royal Society of London* 23(1968), 213–229; "The Accademia del Cimento and Saturn's Ring," *Physis* 12(1970), 36–50; A. F. O'D. Alexander, *The Planet Saturn* (London: Faber and Faber, 1962), pp. 92–100.

3. At the time Maxwell wrote his essay for the Adams Prize, some British writers attributed this discovery to William Ball (1656), but it was subsequently established that this was based on a misunderstanding of Ball's observations and that all the credit should go to Cassini. See Alexander, op. cit. (note 2), p. 115. According to Daniel Kirkwood, the division is due to resonance perturbations by the satellites of Saturn, especially Mimas, and is thus analogous to "Kirkwood gaps" in the asteroid belts. See D. Kirkwood, "On the nebular hypothesis, and the approximate commensurability of the planetary periods," *Monthly Notices of the Royal Astronomical Society* 29(1869), 96–102. For an alternative cosmogonical explanation see H. Alfvén and G. Arrhenius, *Evolution of the Solar System* (Washington, DC: National Aeronautics and Space Administration, 1976), pp. 301–314.

4. Alexander, op. cit. (note 2), p. 119; Albert Van Helden, private communication to S. G. B., 25 October 1976.

5. Immanuel Kant, *Universal Natural History and Theory of the Heavens*, translated by W. Hastie (1900), reprinted with a new introduction by M. K. Munitz (Ann Arbor; University of Michigan Press, 1969), pp. 113–125. The first German edition was printed in 1755.

6. Laplace, "Mémoire sur la theorie de l'anneau de Saturne," *Mémoires de l'Academie royale des sciences de Paris* (1787), published 1789, reprinted in *Oeuvres complètes de Laplace* (Paris: Gauthier-Villars, 1878–1912), vol. 11, pp. 275–292.

7. The rings are "solides irreguliers d'une largeur inégale dans les divers points de leur circonferences, en sorte que leurs centres de gravité ne coincident point avec leurs centres de figure. Ces centres de gravité peuvent être considérés comme autant de satellites qui se meuvent autour du centre de Saturne à des distances dépendantes de l'inegalité des parties de chaque anneau et avec des vitesses de rotation égales à celles de leurs anneaux respectifs" (*Oeuvres*, vol. 11, p. 291). The theory is given in detail in Laplace's *Mécanique Céleste*, vol. 2 (Paris, 1799), livre III, chapitre VI. There are additional helpful notes in the Bowditch translation, *Celestial Mechanics*, vol. 2 (New York: Chelsea, reprint 1966 of 1832 edition), pp. 492–518. Further discussion is contained in Isaac Todhunter, *A History of the Mathematical Theory of Attraction and the Figure of the Earth*, vol. 2 (New York: Dover, reprint 1962 of 1873 edition), pp. 65–73. In the same section Todhunter also discusses Plana's ideas.

8. Laplace, *The System of the World*, translated by H. H. Harte (London: Longmans Green, 1830), pp. 360–361 (from the fifth French edition, 1824).

9. Laplace, *Mécanique Céleste*, tome 5, op cit. (note 7), livre XIV, chapitre III.

10. Alexander, *The Planet Saturn*, p. 133. John Herschel elaborated Laplace's theory in his *Outlines of Astronomy*, in a section that was retained in editions printed long after the idea of solid rings had been abandoned; for example, see the sixth edition, 1859, pp. 346–349. There is an interesting (though amateurish) critique of the Laplace-Herschel theory by James Elliott, a mathematics teacher in Edinburgh, together with a description of an original mechanical model for Saturn's ring, in a prize essay read to the Royal Scottish Society of Arts in 1854. See James Elliott, "A

description of certain mechanical illustrations of the Planetary Motions, accompanied by theoretical investigations relating to them, and, in particular, a new explanation of the stability of the equilibrium of Saturn's Ring," *Edinburgh New Philosophical Journal* 1(1855), 310–335.

11. Johann H. Schroeter, *Kronographische Fragmente zur genauern Kenntniss des Planeten Saturn, seines Ringes und seiner Trabanten* (Göttingen: In Commission der Vandehök-Ruprechtischen Buchhandlung, 1808), pp. xx, 179–180. John Robison, *A System of Mechanical Philosophy*, with notes by David Brewster (Edingburgh: J. Murray, 1822), vol. III, pp. 264–267. G. Plana criticized Laplace's argument that the ring is divided into several concentric rings, but did not give an alternative theory. See his letter in *Correspondance Astronomique, Geographique, Hydrographique et Statistique* 1(1818), 346–350. He published a detailed treatment in "Solution de différents problèmes relatives à la loi résultante de l'attraction exercée sur un point materiel par le cercle etc.," *Memorie della Reale Accademia delle Scienze di Torino* 24(1820), 389–450. According to G. P. Bond (1851 paper cited in note 16), Plana concluded that "more than one ring is not essential," but "the data which he assumed we now know to have been very wide of the truth, as regards the mass and thickness of the ring."

12. See Bessel's views, quoted by P. Volpicelli, "Sur la nature probable des anneaux de Saturne et sur le bolide signalé le 31 août aux environs de Rome," *Compt. Rend. Acad. Sci. Paris* 75(1872), 954–955.

13. E. Roche, "Mémoire sur la figure d'une masse fluide, soumise à l'attraction d'un point éloigné," *Mémoires de la Section des Sciences, Academie des Sciences et Lettres de Montpellier* 1(1849), 243–262; 1(1850), 333–348; 2(1851), 21–32; the remark on Saturn's rings is in 1(1849), 258. In a later paper Roche gave an extended discussion and noted that his first result had been announced before those of Maxwell and Hirn; see "Essai sur la constitution et l'origine du système solaire," ibid. 8(1873), 235–324, especially pp. 292–295.

14. Alexander, *The Planet Saturn*, pp. 166–180.

15. Ibid., pp. 181–183.

16. G. P. Bond, "On the Rings of Saturn," *American Journal of Science* [2] 12(1851), 97–105; also in *Astronomical Journal* 2(1851), 5–8, 10; Edward S. Holden, *Memorials of William Cranch Bond ... and of his son George Phillips Bond* (San Francisco: Murdock, 1897), pp. 254–255.

17. B. Peirce, "On the Constitution of Saturn's Ring," *American Journal of Science* [2] 12(1851), 106–108; Holden, *Memorials*, p. 255.

18. D. Kirkwood, "On Saturn's Ring," *American Journal of Science* [2] 12(1851), 109–110. On Kirkwood's contributions to the nebular hypothesis, see R. L. Numbers, "The American Kepler: Daniel Kirkwood and his analogy," *Journal of the History of Astronomy* 4(1973), 13–21; Kirkwood, "On the Nebular Hypothesis," *American Journal of Science* [2] 30(1860), 161–181.

19. Otto Struve, "Sur les Dimensions des Anneaux de Saturne," *Recueil de Mémoires presentés a l'Academie des Sciences par les Astrononomes de Poulkova* 1(1853), 349–385. The quotation is from the abstract in *Monthy Notices of the Royal Astronomical Society* 13(1853), 22–24.

20. J. R. Hind, "On Saturn's Rings," *Monthly Notices of the Royal Astronomical Society* 15(1854), 31–33; W. S. Jacob, "Remarks on Saturn as seen with the Madras Equatorial ... ," ibid. 13(1853), 240–241.

21. Angelo Secchi, "On the rings of Saturn," *Monthly Notices of the Royal Astronomical Society* 15(1856), 50–56. F. Kaiser, "De Stelling van Otto Struve, omtrent het breeder worden van den Ring van Saturnus, getoetst aan de handschriften van Juygens en de naauwkeurigheid der latere waarnemingen," *Verslagen en Mededeelingen der K. Akademie van Wetenschappen, Amsterdam* 3(1855), 186–232—abstract in *Monthly Notices of the Royal Astronomical Society* 15(1856), 66–72; William Lassell, "Observations of two of the satellites of Saturn and of the exterior diameter of the exterior ring made at Valetta," *Monthly Notices* 13(1853), 181–183; Alexander, *The Planet Saturn*, pp. 184, 336; Agnes M. Clerke, *A Popular History of Astronomy during the Nineteenth Century*, 4th ed. (London: A. and C. Black, 1902), p. 301.

22. For further details and references see the article on Adams by M. Grosser in *Dictionary of Scientific Biography* (New York: Scribner, 1970) 14 vols., vol. 1, pp. 53–54. It is ironic that James Challis, who is generally seen as having frustrated Adams's work on Neptune by not looking for the planet at the time and place when it could be seen, was later in a position to control the administration of the Adams Prize.

23. S. Parkinson was senior wrangler in 1848, the year William Thomson was second wrangler. In 1856 Parkinson was a Fellow of St. John's College.

24. George Biddell Airy (1802–1892) was already the Astronomer Royal. His forte was reducing large amounts of data and introducing rapid methods of calculation for doing so.

25. Challis to Thomson, February 28, 1855 Cambridge University Library, Kelvin MSS, Add. 7618 Box 3.

26. Challis to Thomson, March 14, 1855, same collection.

27. Challis to Thomson, March 23, 1855, same collection.

28. "*The Motions of Saturn's Rings*. The Problem may be treated on the supposition that the system of Rings is exactly, or very approximately, concentric with Saturn, and symmetrically disposed about the plane of his equator, and different hypothese may be made respecting the physical constitution of the Rings. It may be supposed (1) that they are rigid; (2) that they are fluid, or in part aeriform; (3) that they consist of masses of matter not mutually coherent. The question will be considered to be answered by ascertaining, on these hypotheses severally, whether the conditions of mechanical stability are satisfied by the mutual attractions and motions of the Planet and the Rings.

"It is desirable that an attempt should also be made to determine on which of the above hypotheses the appearances of both the bright rings and the recently discovered dark ring may be most satisfactorily explained; and to indicate any causes to which a change of form, such as is supposed from a comparison of modern with earlier observations to have taken place, may be attributed": *The Scientific Papers of James Clerk Maxwell*, ed. W. D. Niven, 2 vols. bound as 1 (New York, Dover reprint, 1965), vol. 1, p. 288.

29. Maxwell, "On the Theory of Rolling Curves," *Transactions Royal Society of Edinburgh* 16(1849), 519–540, and "On the Description of Oval Curves and those having a plurality of Foci," *Proceedings Royal Society of Edinburgh* 2(1857), 89–91. Both papers are reprinted in *Scientific Papers*, vol. 1, pp. 4–29 and 1–3, respectively.

30. Norton Wise, "The mutual embrace of electricity and magnetism," *Science* 203(1979), 1310–1318.

31. A. T. Fuller, "Clerk-Maxwell's London Notebooks: Extracts Relating to Control and Stability," *International Journal of Control* 30(1979), 729–744, repeats the story

III
Borelli's model used by the Accademia del Cimento to test Christian Huygens's ring hypothesis. From Huygens's *Oeuvres complètes*, 22 vols. (The Hague: Société Hollandaise des Sciences, 1888–1950), vol. 3, p. 154.

IV

Saturn as pictured by John Herschel. From the new edition of his *Outlines of Astronomy* (Philadelphia: Blanchard and Lea, 1853).

1853 : Nov. 2.   (*Dawes.*)

1848.   (*W. C. Bond.*)

1856 : Jan. 8.   (*Jacob.*)

V
Diagrams of Saturn and its ring system drawn by Dawes, Bond, and Jacob. From
G. F. Chambers's *Descriptive Astronomy* (Oxford: Oxford University Press, 1867),
book I, plate vi.

VI, VII, VIII

Saturn. From R. A. Proctor's *Saturn and Its System* (London: Chatto and Windus, 1882), plates I and IX.

VI

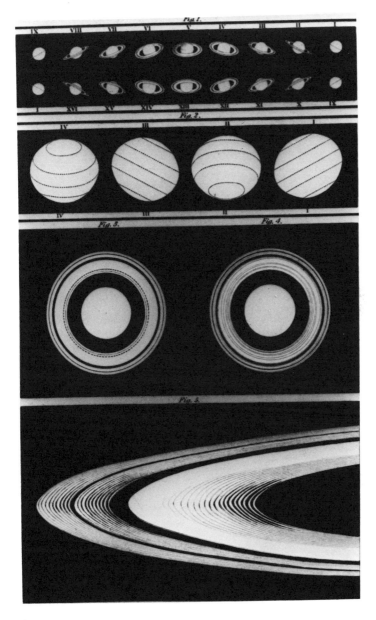

VII

VIII

Documents

1. Letter from Maxwell to R. B. Litchfield, July 4, 1856[a]

Trinity College, Cambridge, Add. MS Letters C 1[83]; quoted in part in
Lewis Campbell and William Garnett, *The Life of James Clerk Maxwell*
(London: Macmillan and Co., 1882), p. 261.

Glenlair
July 4, 1856

Dear Litchfield,

Time is going on & I have to arrange matters in advance. I have been
very much engaged here since I came and I expect to have several things
to do after a little so that it is well to see that things will fit together.
The British Asses[b] meet on the 6th of August and close on the 10th. On
the 14th I have to be in Edinburgh at a marriage and thereafter I purpose to go to Aberdeen for a few days and then to Belfast with a cousin
of mine[c] whom I have persuaded to go there to learn engineering and
his future proprietor has asked him and me to come over & see him.
When that is over I will bring my cousin here to stay so long as is convenient and if you or anyone else such as [Vernon] Lushington should
be in Westmorland &c at that time there might be the more company if
you were to turn up.

    I spoke to you but doubtfully about any time I might have after
Cheltenham.[d] I have certainly no time now & I have much more occu-
pation than I expected such as to examine into the state of two sets of
houses & provide wood &c. for roofing them and workmen to do it and
various things of this kind, also to enquire into the merits of the younger
clergy and the sentiments of the parish on the subject, for our minister
died unexpectedly this week and there are no resident proprietors in

the parish except the patron of the living who is a lady of the Romish persuasion who has been for a year in Edinburgh and denies herself to all her friends. At the same time I ought to be thankful, and am, that all the people here stick to their duty both in working and agreeing together and in giving advice when wanted, now that they have lost the experience and the wellfitted plans under which they used to act with confidence.

I have got some prisms and opticals from Edinburgh and am fitting up a compendious colour weaving machine[e] capable of transportation. I have also my top for doing dynamics and several colour-diagrams, so that if I come to Cheltenham I shall not be empty handed. At the same time I should like to hear from you soon.

I have been giving a portion of time to Saturn's Rings, which I find a stiff subject but curious, especially the case of the motion of a fluid ring. The very forces which would tend to divide the ring into great drops or satellites are made by the motion to keep the fluid in a uniform ring.

I find I get fonder of metaphysics and less of calculation continually, and my metaphysics are fast settling into the rigid high style, that is about ten times as far *above* Whewell as Mill is *below* him, or Comte or Macaulay *below* Mill, using above and below convenionally like *up* and *down* in Bradshaw.[f]

Experiment furnishes us with the values of our arbitary constants, but only suggests the form of the functions. Afterwards, when the form is not only recognised but understood scientifically, we find that it rests on precisely the same foundation as Euclid does, that is, it is simply the contradiction of an absurdity, out of which we may get our legs at last!

a. Richard Buckley Litchfield (1832–1903) was a college friend of Maxwell. His wife Henrietta (daughter of Charles Darwin) wrote his biography (London, 1903).

b. Members of the British Association for the Advancement of Science.

c. William Dyce Cay.

d. Maxwell was setting mathematics examinations at Cheltenham College.

e. This refers to Maxwell's portable color box. See Maxwell," On the Theory of Compound Colours and the Relations of the Colours of the Spectrum," *Phil Trans.* (1860) 57–84 reprinted in *Scientific Papers*, vol. 1, pp. 410–444.

f. In Bradshaw's railway timetable the direction *toward* London is defined as *up* and the direction *from* London is *down*.

2. Letter from Maxwell to William Thomson, August 1, 1857

Glasgow University Library, Kelvin Papers, M6.

Glenlair
Springholm, Dumfries
1st August 1857

Dear Thomson

I have been brewing at Saturn's Rings with infusion of your letters for a month during most of which I have been on the move but I hope to explain myself now. I have had talks with Challis and hunts in the University Library and sights of diagrams. As for the rigid ring I ought first to speak of Prof. Peirce. He communicated a large mathematical paper to the American Academy on the Constitution of Ring One [?] ⟨July 18, 1851⟩ but up to the present year he has no intention of publishing it. There is a "popular" abstract of it in Goulds Astron. journ. June 16, 1851.[a] His result is about 20 fluid rings unable to preserve themselves but guarded by the satellites in some unexplained way.

To understand him it is best to refer to a paper "On the Adams Prize Problem of 1856" in Goulds journal Sept. 5, 1855.[b]

The number of rings is found on the equilibrium theory by conceiving the condition of the surface of a surface of a section of the ⟨the⟩ each ring.

Then the continuity of each ring is disposed of by asserting the tendency of the fluid pressures to equalize the section all round. Now the fluid pressures are those due to the attraction of the ring on itself towards the middle of its section and these are exceedingly small compared with the longitudinal attractions. In fact it is impossible for an elongated fluid mass to be stable without a solid core, much denser than the fluid. (The condition is that the potential at the surface of the fluid due to the fluid and core, together, shall be less as the fluid is thicker.)

It is only dynamically that stability can exist, and it appears from my solution that if there had been statical stability it would have been dynamically unstable.

Peirce also tacitly assumes that the character of the motion of such rings is "steady" and deduces from ⟨that⟩ the equality of attraction on all sides of the planet, and therefore, he says, if the centre of the planet were in motion with respect to the ring, it would continue to move till it met the ring as the attractions of the parts of the ring always balance by reason of the nearer and faster moving parts being thinner.

But he forgets that *during* these changes this law of thickness is so much modified that the reasoning fails, and besides there is nothing but a mathematical assumption to preserve the "steady motion" of the rings all the time. With such an assumption he finds it easy to convert the rings into a comet at aphelion by becoming more and more elliptic.

Now for the body guard of Satellites.

It is necessary that the centres of Saturn & the Rings should keep together. According to Peirce there is no reason why they should as far as they are concerned. But the Satellites act on both and pretty much the same way so that they cause them to perform pretty nearly the same paths and *so* to keep pretty nearly together.

I hope I state the argument right for it is greatly abridged in the journal.

His plan of drawing the potential surfaces for the solid ring is very good. I intend to do it actually on paper as I have done electric lines. But I strongly suspect the dynamics of his reasoning on the stability. The motion (in the plane of the ring) depends on 3 variables, 2 for the centre of gravity and one for rotation. All these are interdependent so that an oscillation in one affects the rest. Now the man who without previous analysis in the ordinary way, can so conceive the motion as to predict the whole effects of a disturbance, and then describe in ordinary language the nature of his reasoning, has no business to contribute to mathematical journals and such like. He ought to sit down at once and write off a new Principia on Spiral Nebulae. When I have carefully worked out the motion I hope to state its nature and give a rough description of the sequence of events but that is not done yet.

I must get up the question put by you and Challis about the stability of a ring nearly uniform.[c] That is also not done nor do I relish it. Now for the streams of particles or of fluid. The equations will do if mK is not too large. Now K is inconceivably small in fact the smaller the better. m is the number of undulations round the ring. If m be very great the supposition, that the undulations were long with respect to the breadth of the ring will not be true, and the assumed value of K will be too large.

The calculation in this case is too difficult for me but I think that mK cannot exceed a certain value which depends on the section and density of the ring, so that a broad ring would stand where a thin one would snap. If you admit the fluid rings with internal friction in each ring, these short waves disappear most rapidly and are therefore less to be dreaded.

If you take the discrete particle theory with a linear ring of them, still K decreases as m increases.

Also consider the smallness and regularity of all disturbance by satellites as long as Saturn is so large and they so small.

What shall we say to a great stratum of rubbish jostling and jumbling round Saturn without hope of rest or agreement in itself till it falls piecemeal and grinds a fiery ring round Saturns equator, leaving a wide tract of lava with dust and blocks [?] on each side and the western side of every hill battered with hot rocks?

First then there is no way of making a visible thing without refraction except of discrete parts with interstices. That is the cheapest way of attenuating comets and other heavenly bodies. Gas is quite a mistake. The bigger the bits the more transparent at the density. Nebulae go naturally in bits, why not rings. Now the edge of Saturn is seen quite undisturbed by the dark ring.

Next what will be the effect of this stratum not disposed in thin rings as before but spread over an annular surface of great breadth and moving with various velocities.

The calculation of the undulations due to attraction is far beyond me. We have two dimensions for one and all the spreading effects at the different velocities.

What will be the effect of collisions. 1st They destroy vis viva but not angular momentum so that if the orbits remain circular they spread the ring wider on the whole tending to make a division in it. 2nd Suppose that disturbance produces ellipticity in an orbit of a particle so that it meets a particle belonging to a wider orbit. The first particle at the most distant part of its course travels slower than the second so that the first is kicked forward and the second backward. The ⟨inner particle is brought to a more distant orbit on⟩ The mean distance of the inner particle is increased and that of the outer diminished by this collision and both orbits become less eccentric.

This is the opposite effect, tending to concentrate the ring again.

It will be seen that this effect tells most on the inner and outer edges of the whole system, so that these will be held in while the rings generally are spreading. Now the obscure ring is sensibly thicker at its inner edge. As for the outer rings they are far too massy in proportion to the inner ones for any visible result to take place on them. The general result is therefore rather a spreading of the mean parts of the rings towards the outer and inner edges, than an enlargement of the whole ring, but in the inside the spreading effect may be so much the stronger as the stuff spread is thinner.

The best historical treatise is in Recueil de Memoires Astronomiques de Poulkova—Sur les dimensions des Anneaux de Saturn Par M. Otto Struve 14 Nov. 1851.[d]

The encroachment of the inner ring seems very certain and not due to the improvement of telescopes.

On the whole I should not recommend any one to feu[e] a building stance on any of the Rings without security that his parallelograms may not be spun out into spirals of unknown extent in a few hours. Arctic ice packs are most secure in comparison. As for the men of Saturn I should recommend them to go by tunnel when they cross the "line."

Yours truly

J. C. Maxwell

a. Benjamin Peirce (1809–1880), American mathematician and theoretical astronomer, professor at Harvard University. His paper "On the Constitution of Saturn's Ring," *American Journal of Science* 12(1851), 106–108, *Astronomical Journal* 2(1851–52), 17–19, and *Annalen der Physik* 84(1851), 313–319, is the one referred to here.

b. B. Peirce, "On the Adams Prize-Problem for 1856," *Astronomical Journal* 4(1856), 110–112. The editor of this journal was Benjamin Apthorp Gould (1824–1896), who had founded the journal in 1849. The continuation Peirce promised at the end of his paper did not appear, at least not under the same title.

c. As noted in the introduction James Challis and William Thomson were the examiners for the Adams Prize competition; see the published statement of the problem in note 28 to the introduction. Maxwell's original version of his essay had been submitted in late 1856, assuming that it is the one metioned by Challis in his letter to Thomson, December 31, 1856.

d. See the introduction, note 19. On the basis of his review of observations going back to the seventeenth century, Struve claimed that the ring was spreading, and this claim was one of the main reasons for the choice of the topic for the Adams Prize.

e. This is a Scottish legal word, a variant of fee. A feu is a perpetual lease at fixed rent, or a piece of land so held.

3. Letter from Maxwell to William Thomson, August 24, 1857

Glasgow University Library, Kelvin Papers, M7.

Glenlair
24th August 1857

Dear Thomson

Since I last wrote to you I have wrought at the theory of a solid ring. I had made a mistake in the evaluation of the quantities $d^2 V/dr^2$, $d^2 V/d\varphi^2$. I ought to have treated them by the condition

$$\frac{d^2 V}{dr^2} + \frac{1}{r}\frac{dV}{dr} + \frac{1}{r^2}\frac{d^2 V}{d\varphi^2} + \frac{d^2 V}{dz^2} = 0$$

where $d^2 V/dz^2$ refers to a direction perp. [perpendicular] to plane of $r\,d\varphi$. In this way I find

$$\frac{d^2 V}{dr^2} = \frac{R}{2a^3}(1+g) \quad \left|\frac{d^2 V}{dr\,d\varphi}\right| = \frac{R}{2a^2}fh \quad \frac{d^2 V}{d\varphi^2} = \frac{R}{2a}f^2(3-g)$$

where f is the ratio of the distance of centre of gravity of ring from centre to radius and g & h depend on the form of the ring. The density to unit of length being of the form

$$\mu = \frac{R}{2\pi a}\left\{1 + 2f\cos\theta + \frac{2}{3}g\cos 2\theta + \frac{2}{3}h\sin 2\theta + \&c\right\}$$

The general differential equation of motion becomes

$$(1 - f^2)\frac{n^4}{\omega^4} + \left(1 - \frac{5}{2}f^2 + \frac{1}{2}f^2 g\right)\frac{n^2}{\omega^2} + \frac{9}{4} - 6f^2 - \frac{1}{4}(g^2 + h^2)$$

$$+ 2f^2 g = 0$$

$$A\frac{n^4}{\omega^4} + B\frac{n^2}{\omega^2} + C = 0$$

and this equation must give two real negative values of n^2 which implies that A B & C must have the same sign and that $B^2 > 4AC$.

Now if we put $g = 0$, $h = 0$ the first condition gives $f^2 < 3.75$ and the second $f^2 > 37445$.[a] But in order that the section of the ring may be real throughout (no *negative mass* in it) we must have $f < \frac{1}{2}$ so that we are blocked out.

Now try a uniform ring, mass $= Q$ with a heavy particle P at a point in its circumference

$$P + Q = R \qquad f = \frac{P}{R} \qquad g = \frac{3P}{R} = 3f$$

The first condition gives $f < .8274$
The second $f > .815865$

Between these values there is stability and there is nothing to hinder the mass of P to be to that of Q as 82 to 18.
   Now take $f = 82$ and we find

$$\sqrt{-1}\frac{n}{\omega} = \pm .5916 \qquad \text{or} \qquad \pm 3076$$

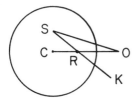

[Diagram 1]

indicating periods of 1.69 or 3.25 revolutions. Working out the whole motion we find[b]

$$\frac{r_1}{a} = A\sin(.59\theta_0 - \alpha) + B\sin(.3\theta_0 - \beta)$$

$$\theta_1 = 3.21A\cos(.59\theta_0 - \alpha) + 5.72B\cos(.3\theta_0 - \beta)$$

$$\varphi_1 = -1.2A\cos(.59\theta_0 - \alpha) - .79B\cos(.3\theta_0 - \beta)$$

Here are 3 variables $r_1\ \theta_1\ \varphi_1$ expressed with 4 arbitraries $A\ B\ \alpha\ \beta$. Since $r_1\ \theta_1\ \varphi_1$ & their diffls [differentials] $dr_1/dt\ d\theta_1/dt\ d\varphi_1/dt$ are all disposable we should expect 6 arbitraries. But two of them are swallowed up by assuming r_0 & θ_0 as the mean values of r & θ, and we prove the thing analytically *by searching for dropped roots* in the equation. It is easy to show [diagram 1] that for points near C the centre of the ring in this case the attractive force is directed always towards a point O, further from C than R the centre of gravity. Now how is the ring to revolve round S (Saturn) if the force acts at a point beyond the centre of gravity?

Suppose the line RO in advance of SR the force in OS tends to increase the rotation of the ring and so to ⟨bring⟩ increase KRO the angle of libration.

But in order to estimate the whole effect we must consider the force resolved into a couple acting as aforesaid and radial & tangential forces on R of which the tangential will be negative and by its *indirect* effect will *increase* the angular velocity of SR.

Whether SR will *overtake RO* must depend on numerical considerations. In that case there is *stability and periodic inequality*.

I have tested the whole of this part well and found it all right. Next comes the fluid filament. Now if you or anybody else can tell me what ⸕ is the attraction towards the centre upon a particle in the centre of the transverse section of a ring whose mean radius is a, radius of section b (b small) and density ρ I will be greatly obliged for I am very bad at such calculations.

[Diagram 2]

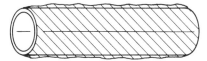

[Diagram 3]

Or if a solid ring of these dimensions were cut in two [diagram 2] and the parts placed in contact what would be the whole pressure between them? I am more entangled with questions of this kind than with anything else.

Then here is another nice problem a right cylinder radius $= b$ has electricity (or anything else) spread over its surface according to the law

$$\rho = A \cos \frac{r}{2\pi C}$$

find the potential at any point.

Or solve the eq^n

$$\frac{d^2 u}{dr^2} + \frac{1}{r} \frac{du}{dv} = n^2 u = 0$$

To return to visibles & tangibles [diagram 3]—Let there be an infinitely long cylinder of any substance, and let there be spread upon it a coating of another substance to the thickness

$$T = A + B\cos \frac{r}{2\pi C}$$

where A and B are very small, so as to be in waves of thickness & thinness along the cylinder then conceive this coating melted suddenly. Will the irregularities increase or diminish owing to the attractions of the mass?

If C, the wave length is *small* compared with the diameter of the cylinder they will diminish. If it is *large* they will increase. What is the critical value of C?

If the force were capillary tension on the surface it is easy to show that $C = 2\pi b =$ circumference of cylinder at critical period, and

Plateau's experiments show that c/b is between 3 & 3.5.[c] But the same problem in *attraction* is beyond my calculating power.

My theory of *long waves* is all right. The longitudinal force is toward the thicker parts and the transverse force is similar to that due to a tension increasing with the thickness.

But if the waves are *short* the tendency is toward the thinner parts, while the transverse force is still of opposite sign to the transverse displacement but depending more on *displacement* ⟨than⟩ from the mean position than on *curvature* at the point.

The case is like that of long and short waves of light. The attractions are not confined to contiguous particles, so that though long waves may be treated so, short ones cannot, so that the theory of long waves is not applicable to them.

I intend to apply what I know of this to the fluid rings as I go along. The last sentence applies equally to the hailstorm ring, and my scoffing allusion to the *circular* [?] was on account of my not having anything ready at the time to explain it completely. I intend also to accumulate notions about the hailstorm ring and to give the explanation of M[r] Elliott's experiment of the iron ring spinning round a magnet.[d] That is a problem in 3 dimensions and it depends on the plane of the ring having to turn ⟨in its⟩ when its centre moves, owing to there being a fluid point connected with the ring, out of its own plane.

Next Monday my address will be at
John MacCunn Esq[re]
Ardhallow
Dunoon.[e]
After Monday 7th Sept., I shall be at Roshven,[f] Strontian, for a while after which I hope to be here again with Mrs. Wedderburn for the rest of my time.

Yours truly
J. C. Maxwell

a. In Maxwell, "On the stability of the motion of Saturn's Rings," *Scientific Papers*, vol. 1, p. 307, reproduced in document 17, where these numbers are given as 0.375 and 0.37445. Presumably Maxwell discovered a factor-of-ten error and dropped the decimal point in the second number in this letter.

b. See equations (40)–(42) in Maxwell, "Saturn's Rings," *Scientific Papers*, vol. 1, p. 309, where most of the numerical coefficients are similar (though given to four significant figures) except for the coefficient of B in the third equation, which is 5.79 in place of the 0.79 of this letter.

c. Joseph Antoine Plateau (1801–1883) was a Belgian physicist. He was well-known in the nineteenth century for his experiments on the effects of rotation on a sphere of oil immersed in a mixture of alcohol and water whose density was equal to that of the oil. He was able to reproduce the formation of rings and satellites in a manner strikingly reminiscent of Laplace's nebular hypothesis. See Maxwell, "Saturn's Rings," *Scientific Papers*, vol. 1, p. 396.

d. James Eliott, a mathematics teacher in Edinburgh, presented a paper on his model for Saturn's ring's to the Royal Scottish Society of Arts, February 27 and March 13, 1854; it was awarded a silver medal. See Eliott, "A description of certain mechanical illustrations of the Planetary Motions, accompanied by theoretical investigations relating to them, and, in particular, a new explanation of the stability of equilibrium of Saturn's Rings," *Transactions of the Scottish Society of Arts, Edinburgh* 4(1856), 318–344, and in the *Edinburgh New Philosophical Journal* 1(1855), 310–335.

e. John MacCunn was the husband of Mrs. Maxwell's sister.

f. The residence of Maxwell's cousin Jemina and her husband Hugh Blackburn, professor of mathematics at Glasgow University.

4. Letter from Maxwell to Lewis Campbell, August 28, 1857

Lewis Campbell and William Garnett, *The Life of James Clerk Maxwell* (London: Macmillan and Co., 1882), p. 278.

Glenlair
28th, August 1857

... I have been battering away at Saturn, returning to the charge every now and then. I have effected several breaches in the solid ring, and now I am splash into the fluid one, amid a clash of symbols truly astounding. When I reappear it will be in the dusky ring, which is something like the state of the air supposing the siege of Sebastopol[a] conducted from a forest of guns 100 miles one way, and 30,000 miles the other, and the shot never to stop, but go spinning away round in a circle, radius 170,000 miles . . . .

a. The eleven-month siege of Sebastopol by British, French, and Turkish forces was the dominant event of the Crimean War. When it ended on September 11, 1855, the city had been virtually reduced to a heap of rubble. Leo Tolstoy's account of the siege from the Russian side is to be found in *The Cossacks* (1863).

5. Letter from Maxwell to R. B. Litchfield, October 15, 1857

Trinity College, Cambridge, Add. MS Letters C 1[88]; quoted in part in Lewis Campbell and William Garnett, *The Life of James Clerk Maxwell* (London: Macmillan and Co., 1882), pp. 282–283.

Glenlair
15th Oct. 1857

Dear Litchfield

I enclose what Sale sent me.[a] We are all losing our friends. My aunt Mrs. Wedderburn who is with me has for the last 3 months been gradually learning the details of the murders of her cousin John Wedderburn and his wife & child. Even now there are manifestly contradictory reports by Englishmen of what they have seen or done. Her own son John & wife are at Moulton where they have had to disarm the troops and to dispatch 9 of those who remained with them & professed friendship but were traitors. She is 68, and after four years of constant pain of body and torture of mind she has completely recovered health and strength and comfort. It is fearful to think what may happen to one through bodily infirmity, so that while appearing not unwell there may be a dull nameless pressure on the mind relieved only by bodily pain or unreasonable mental distress of a more acute kind, but all this is away since this time last year, and now she can bear anything and talk about anything, even the darkness that is past.

I was glad Sale sent me the letter. Remember that besides all the danger and distance from friends, there was the fever, of which he[b] had already long experience, and which in such times he knew to be as inconvenient to his friends as himself. But it is no use saying this and that. Some men redeem their characters by their deeds, and we praise them. Those that merely show their character by their deeds should be remembered, not praised; and a complete true man will live longest in the memory, and I cannot but think will be less changed in reality, than one who has doubtfully struggled with duplicity in his constitution, and has walked with hesitation, though along a good path. I know that both do deny and renounce themselves in favour of duty and truth, as they come to see them; and as they come to see how goodness, having the knowledge of evil, has passed through sorrow to the highest state of all, they accept it as a token that they have found their true head and leader, and so, with their eyes on him, they complete the process called the knowledge of Good and Evil, which they commenced so early and so ignorantly.

Now what the "completion of the process" is I cannot conceive, but I feel the difference of Good and Evil in some degree, and I can conceive the perception of that difference to grow by contemplating the Good till the confusion of the two becomes an impossibility. Then comes the mystery. I have memory and a history, or I am nothing at all. That memory and history contain evil, which I renounce, and must still maintain that I was evil. But it contains the image of absolute good, and the fight for it, and the consciousness that all this is right.

So there the matter lies, a problem certain of solution.

I am & have been very busy here. I leave this on 19th. Address Lauriston Lodge, Edinburgh. After 26th—Marischal College Aberdeen. I am grinding hard at Saturn and have picked many holes in him and am fitting him up new and true. I am sure of most of him now, and have got over some stumbling blocks which kept me niggling at calculations 2 years.

I am to have some artisans as weekly students this winter.

I was with Wigram M. P. Camb. [Cambridge] lately and discussed University matters a little. His wisdom was not equalled by his information, but the sense was good. It is awfully late and Toby[c] always makes a row when he goes to bed so I must shut you up as I have more to write.

I am so glad you are glad after you have seen Farrar that I will not even demand a local habitation or a name for the twin soul.[d]

But Lewis Campbell (a much older friend) I believe to be nearly as glorious since January[e] and my people had an actual visé the old ones and the youngest ones conjointly & severally & the old gune grad[f] votes down to a young lady of 6 years who has had letters from several young men already & ought to know all about it.

We are still in our domestic troubles. The sister of the little girl that died is just breathing for the last few days and there are bad symptoms but not hopeless. A very little one is very ill and several of other of our people have been ill and in a slighter way than in that family. So we wait.

Yours truly

J. C. Maxwell

a. In the first paragraph of this letter Maxwell is reporting on deaths in his family during the Indian Mutiny of 1857.

b. Here Maxwell is trying to come to terms with the death of a close college friend, Robert Henry Pomeroy (1832–1857), also during the Indian mutiny. Maxwell returns to this theme of death and the loss of friends in a later letter to Litchfield, February 7, 1858.

c. Toby was Maxwell's dog.

d. Frederick William Farrar (1831–1903), the future dean of Canterbury, was a college friend of Maxwell. See Campbell and Garnett, *Life* (1882), chapter VI, and pp. 618–621. The twin soul is presumably Lucy Mary Cardew, whom Farrar married in 1860.

e. For Lewis Campbell's glorious state see *Life*, pp. 264–266.

f. gune grad: presumably "married state"—from the Greek γυνη, woman.

6. Letter from Maxwell to H. R. Droop, November 14, 1857[a]

Lewis Campbell and William Garnett, *The Life of James Clerk Maxwell* (London: Macmillan and Co., 1882), p. 291.

129, Union Street, Aberdeen
14th November 1857

I am very busy with Saturn on the top of my regular work. He is all modelled and recast, but I have more to do to him yet, for I wish to redeem the character of mathematicians, and make it intellegible. I have a large advanced class for Newton, physical astronomy, the electrical sciences and optics. What is your department by the way?

I have also a mechanics class in the evening, once a week, on mechanical principles, such as the doctrine of the lever, work done by machines etc. So I have 15 hours a week which is a deal of talking straight forward.

I am getting several tops (like the one I had at Cambridge) made here for various parties who teach rigid dynamics.[b]

a. H. R. Droop (1832–1884) was a Cambridge friend and third wrangler in 1854.

b. The dynamical top referred to is one Maxwell designed and had built to exhibit the motion of the earth. It is described in Maxwell, "On a Dynamical Top, for exhibiting the phenomena of the motions of a body of invariable form about a fixed point, with some suggestions as to the Earth's motion," *Transactions, Royal Society of Edinburgh* 21(1857), 559–570; reprinted in *Scientific Papers*, vol. 1, pp. 248–262. The diagram of the top is reproduced from the reverse of the page of diagrams following the paper.

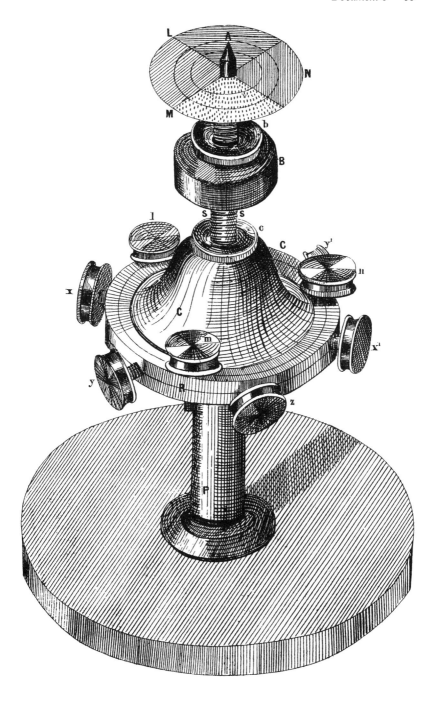

7. Letter from Maxwell to William Thomson, November 14, 1857

Glasgow University Library, Kelvin Papers, M8.

129 Union Street, Aberdeen
14 Nov. 1857

Dear Thomson

I suppose, after a long vacation, you are very busy in the beginning of
the session. Nevertheless you must allow me to report progress on
Saturns rings, so that if you have any remarks to make on my results, I
may have the benefit of them. I have already reported on the rigid rings.
I got your investigation of it, but I have been entirely occupied with the
⟨flui⟩ non-solid rings since so that I have only *read* it, not *worked* it yet.
In the first place, then, I have abolished my off hand theory of the
attractions of a thin fluid filament affected by waves, *long* compared
with the diameter of the filament, and this because the *short* waves are
the only dangerous ones.

I have therefore *begun* with the theory of a ring consisting of μ satellites affected with undulations of the form $A\cos m\theta$ normal, $B\sin m\theta$
tangential, or rather putting ρ & σ for ⟨nor⟩ radial & tangential displacements & s for θ

$$\rho = A\cos(ms + nt + \alpha)$$

$$\sigma = B\sin(ms + nt + \alpha)$$

I get a biquadratic for n whose four values are possible, provided Saturn
be large enough. The force on which the value of n chiefly depends is
⟨that ari⟩ the tangential force arising from the attraction of the satellites
towards the crowded parts of the ring. This must never exceed $\frac{1}{14}\omega^2$.
It is greatest when successive particles are in opposite phases, that is,
when $m = \frac{1}{2}\mu$.

In order that there may be stability in this case

$$S > .4352\mu^2 R$$

where S is Saturn, R the mass of ring, μ the number of satellites. If
$\mu = 100$, $S > 4352R$.

If the mass of the ring were too great, then it would run into lumps of
satellites, till the number of satellites was reduced to what the Planet
could govern.

I have traced the relative & absolute motion of each satellite the
instantaneous form of the ring, and the mode of propagation of the

waves both in the case of stability and in the cases where the ring must break up. When the tangential force is *too great* the effect is that all the waves increase in amplitude till the ring gives way. (Case I) but when the tangential force is negative, that is when it tends *from* crowded portions there is *no propagation* but the irregularities increase without limit in the parts of the ring where they originated (Case II). This effect is very remarkable as showing the destructive effect of an apparently conservative force.

I then apply my method to a liquid ring very much flattened in section. I find that the tangential force is positive for long waves and has a maximum when the wave length is 10.353 times the thickness. It is 0 when the wave length is 5.471 times the thickness and is negative for shorter waves.

That Case I of instability may be avoided [but] the density of Saturn must be at least 42.5 times that of the ring. Now that the outer and inner parts should have the same angular velocity Laplace shows that Saturn must not be more than 1.3 times as dense as the ring.[a] Besides, the *very* short waves having a negative tangential effect (from thick to thin) will inevitably produce the case No. 2 of instability so that the liquid continuous ring is doomed.

If a ring were made of a cloud of satellites the condition is that the density of Saturn should be more than 334 times that of the ring.

I am just entering on the case of many rings of satellites mutually disturbing one another which I think I can attack. The general case of a fortuitous concourse of atoms each having its own orbit & eccentricity is a subject above my powers at present, but if you can give me any hint as to the point of attack I will go at it.

Mean while I am busy putting what I have into intelligible form by means of a good deal of verbal explanation and tracing out of the mechanical phenomena.

I also wish to hear from you what you think a good galvanic battery [is] for class purposes and in particular about the expense, convenience or suitableness of the plan of that you mention in your Bakerian lecture.[b] My object is chiefly quantity, to do magnetic experiments well, and to be of use if I set up a great magnet for diamagnetism. I was talking to Prof. Forbes[c] about the Atlantic Telegraph and heard from him that the slipping of the cable down the incline in the direction of its length had given rise to great waste of cable [diagram 1]. In fact the lateral resistance of the water must convert the line of cable into an inclined plane down which it slides and lies in folds below. I hear of buoys &c but would not *kites* be better, as thus—flat pieces of iron or wood fastened at intervals so [diagram 2] [.] The kite to be attached by

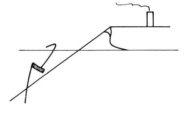

[Diagram 1]

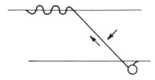
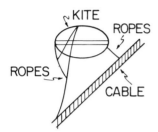

[Diagram 2]

two ropes so that the upper one keeps the plane of the kite rigid, while the lower pulls the rope forwards so as to keep it stretched the whole way as it gradually sinks down to the bottom. The size & frequency of kites to depend on the diameter & weight of the rope and the rate at which it is to be payed out.

Besides keeping the lower part tight the kites would relieve the ⟨force⟩ tension at the surface which is what the public dread, as it really broke the rope, but that was because too much was being spent and the break [brake] has been applied too strong.

My class is small this year. I have viva voce'd them all and will begin written questioning on Tuesday. The second year (my old students[)] have been grinding in the summer and have done well in exam[n] [examinations]. A good many of them come to me as an advanced class.

I have had a correspondence with Faraday on the "Lines of Force" as applied to Gravitation.[d] What a painful amount of modesty he has when he talks about things which may possibly be of a mathematical cast. If I were free to think about these things I think I could put any part of the theory of Gravitation into Faraday's language which I think a most powerful one instead of a mere complication of simple ideas, as some conceive it.

I read your paper on electric qualities of metals with the more pleasure as I knew how you had divided the labour and exercise with your students. I think many people have failed to follow some of the theoretical arguments, for want of faith in reason.[b] I think you might use a little rhetoric with effect, for the satire of such philosophers as the dry spirit repels. I have not heard of your brother James lately. Is he Godwin's successor yet?

I hope Mrs Thomson is able to thrive in Glasgow after her vacation. I met an aunt of hers, Mrs Graham at Dunoon,

Yours truly

J. C. Maxwell

a. See Laplace, *Celestial Mechanics*, translated by Nathaniel Bowditch, vol. II (Boston: 1832; Chelsea reprint, 1966), p. 511.

b. William Thomson's Bakerian Lecture was delivered in 1856. See "On the Electrodynamic Qualities of Metals," *Phil. Trans.* 146(1856), 649–751. This is the paper to which Maxwell refers later in this same letter.

c. James David Forbes, professor of natural philosophy at Edinburgh University.

d. Maxwell's letter to Faraday is reprinted in the second edition of Campbell and Garnett's *Life* (1884) and in the reprint of the first edition (New York: Johnson, 1969). See Robert H. Kargon, preface to the Johnson reprint of Campbell and Garnetts *Life*, pp. xv–xvii. Faraday's reply is in Cambridge University Library, Maxwell MSS.

8. Letter from Maxwell to Peter Guthrie Tait, November 21, 1857

Kings College, Aberdeen, Maxwell Manuscripts.

... I am still grinding at Saturn's Rings. I have shown that any solid ring must be horribly disfigured in order to go at all and in fact dismissed it. A liquid ring will either fly asunder from inequalities of centrifugal force or break into drops by longitudinal forces. *But* the drops as formed may continue as a ring of satellites and will be *spaced out* regularly of themselves provided they be not too large or too numerous for Saturn to govern.

I have found the numerical conditions for the amount of satellites which Saturn can maintain as a ring. How many can rally round him and how many must form coalitions among themselves.

I am now busy with two rings of satellites with different velocities disturbing one another....

9. Letter from Maxwell to Jane Cay, November 28, 1857[a]

Lewis Campbell and William Garnett, *The Life of James Clerk Maxwell* (London: Macmillan and Co., 1882), pp. 292–293.

129, Union Street, Aberdeen
28th Nov. 1857

I had a letter from Willy to-day about jet pumps to be made for real drains, but not saying anything about the Professorship of Engineering.[b]

I have been pretty steady at work since I came. The class is small and not bright, but I am going to give them plenty to do from the first, and I find it a good plan. I have a large attendance of my old pupils, who go on with the higher subjects. This is not part of the College course, so they come merely from choice, and I have begun with the least amusing part of what I intend to give them. Many have been reading in summer, for they did very good papers for me on the old subjects at the beginning of the month. Most of my spare time I have been doing Saturn's Rings which is getting on now, but lately I have had a great many long letters to write,—some to Glenlair, some to private friends, and some all about science. . . . I have had letters from Thomson[c] and Challis[d] about Saturn—from Hayward[e] of Durham University about the brass top, of which he wants one. He says that the Earth has been really found to change its axis regularly in the way I supposed. Faraday[f] has also been writing about his own subjects. I have also to write Forbes[g] a long report on colours, so that for every note I have got I have had to write a couple of sheets in reply, and reporting progress takes a deal of writing and spelling. . . .

I have had two student teas, at which I am becoming expert. I have also indulged in long walks, and have seen more of the country. The evenings are beautiful at this season. There have been some very fine waves on the cliffs south of the Dee.

a. Jane Cay: Maxwell's maternal aunt.

b. Willy was Maxwell's cousin William Dyce Cay, who also studied at Edinburgh University, then went to Belfast to study engineering with James Thomson, the older brother of William Thomson.

c. William Thomson: One of the examiners for the Adams Prize, who contributed an appendix to the published version of Maxwell's essay. See Maxwell, "Saturn's Rings," *Scientific Papers*, vol. 1, pp. 374–376, reproduced in document 17.

d. James Challis: Plumian Professor and director of the Cambridge University Observatory. He first suggested the subject of Saturn's rings for the Adams Prize.

e. Robert Baldwin Hayward (1829–1903) was a Fellow and Assistant Tutor at St. Johns College, Cambridge, from 1852 until he was appointed Tutor and Reader in natural philosophy at the University of Durham in 1855.

f. Michael Faraday (1791–1867) and Maxwell corresponded for a decade about problems in electrostatics, magnetism, and electromagnetism.

g. James David Forbes (1809–1868) was appointed to the chair in natural philosophy at Edinburgh University in 1833. Forbes read Maxwell's first papers to the Royal Society of Edinburgh and allowed Maxwell to work in his own laboratory when Maxwell was a student at Edinburgh. They remained close friends until Forbes's death in 1868.

10. Letter from Maxwell to Lewis Campbell, December 22, 1857[a]

Lewis Campbell and William Garnett, *The Life of James Clerk Maxwell* (London: Macmillan and Co., 1882), pp. 295–296.

129 Union Street, Aberdeen
22d Dec. 1857

... I am still at Saturn's Rings. At present two rings of satellites are disturbing one another. I have devised a machine to exhibit the motions of the satellites in a disturbed ring; and Ramage is making it, for the edification of sensible image worshippers.[b] He has made four new dynamical tops, for various seats of learning.

I have set up a model of Airy's Transit Circle,[c] and described it to my advanced class today. That institution is working well, with a steady attendance of fourteen, who have come of their own accord to do subjects not required by the College, *and the dryest first*.

To the present time we have been on Newton's *Principia* (that is Sects, i. ii. iii, as they are, and a general view of the Lunar Theory, and of the improvements and discoveries founded on such inquiries). Now we go on to Magnetism, which I have not before attempted to explain.

The other class is at two subjects at once. Theoretical and mathematical mechanics is the regular subject, but two days a week we have been doing principles of mechanisms and I think the thing will work well. We now go on to Friction, Elasticity, and Breakage, considered as subjects for experiment, and as we go on we shall take up other experimental subjects germane to the regular course. I am happy in the knowledge of a good tinsmith, in addition to a smith, an optician, and a carpenter. The tinsmith made the Transit Circle...

a. Much of this letter is omitted. The first half consists of Maxwell's thoughts on Lewis Campbell's ordination in December 1857. The last part of the letter contains

Maxwell's account of the proposed merger of Marischal College with Aberdeen. This merger did indeed take place and Maxwell lost his position as professor of natural philosophy at Marischal and he moved to Kings College, London, in 1860.

b. This model was made to illustrate the wave motions generated in a ring of satellites not touching each other. The major force acting on each of the satellites was the gravitational attraction of the main body of Saturn. The model is sketched and its operation explained in the published version of Maxwell's essay (reproduced as document 17).

c. For the transit circle, see George Biddell Airy, "On the Method of Observing and Recording Transits, lately introduced in America," *Monthly Notices of the Royal Astronomical Society* 10(1849–50), 26–34. For a vivid description of Maxwell's model of the transit circle, and its influence on one member of his "advanced class," see D. Gill, *History of the Royal Observatory, Cape of Good Hope* (London: HMSO, 1913), p. xxxi.

11. Letter from Maxwell to William Thomson, January 30, 1859[a]

Glasgow University Library, Kelvin Papers, M10.

129 Union Street, Aberdeen
Jan. 30, 1859[a]

Dear Thomson
I have never thanked you for your propositions on the rotation of a body in a perfect fluid. I hope to be able to give a proper attention to them in the vacation as I am tolerably busy at present. The *rings* have been stationary since New Year. I have had a model of my theory of the waves in a ring of satellites made by Ramage. It is now set up and nearly ready [diagram 1]. It consists of two wheels on the parallel parts of a cranked axle. There are 36 little cranks ranged round the circumference which all move parallel to the axle-crank and to each other. Each carries a satellite placed excentrically on the part of the crank which sticks through the wheel, so that when the axle turns and the wheel moves,

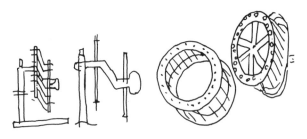

[Diagram 1]

every satellite describes a small circle. The position of the satellite in this circle is arbitary and its arm may be adjusted on the crank at pleasure, so that by properly arranging these arms the ring of satellites may be thrown into waves of any length which travel round the rings, the wheel being fixed.

Fixing the crank and moving the wheel we exhibit the absolute motion of the satellites in the "1st" & "fourth" waves of my essay. The others are less interesting and cannot be well exhibited in any way.

I have been lecturing on statical electricity to the second year, and next week I shall have half a dozen to study "electrical images" over a cup of tea. I begin current electricity on Tuesday. Next year I must set up Daniell properly. I have not had time to think it all over but there seems to be something in this which follows—

Suppose magnetism to consist in the revolution or rotation of any material thing, in the same or the opposite direction to the current of positive electricity which is equivalent to the magnet.

Let an unmagnetized steel bar be fixed to an axis passing perp. [perpendicularly] thru its centre of gravity and let it be mounted in a brass frame so adjusted that the moments of inertia in the plane perp. [perpendicular] to the axis are nearly equal but yet that round the steel bar decidedly the least [diagram 2].

Let this apparatus be placed in a revolving frame, so that the axle of the magnet is perp. to that of the frame. If the axle of the frame be placed magnetic east and west, the axle of the magnet *AB* will revolve in the magnetic meridian. Now, when all is balanced, let the steel bar be made a magnet (NS). The Earth will set it perp. to the axle E.W. Now let it be set in rotation about EW. Because of the moment of inertia about NS is less than that about the perp. axis, the motion will

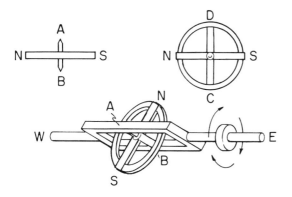

[Diagram 2]

be stable (by centrifugal force) when NS is perp. to EW. But the brass frame may be so adjusted that this stability is very small, so that a slight force will produce a deviation of sensible magnitude.

Let the axis EW revolve in the direction of the arrows then if any thing in the magnet is in a state of rotation, the axis of its rotation will tend to be in the same direction as EW so that if N be the end which *points to the North* and if the rotation be in the direction of the *positive* current, N will go towards W and from E.

Now let everything be the same and let the magnet be reversed by remagnetising it without taking it out, then it ought to deviate the other way. Also the deviation ought to be reversed by reversing the direction of rotation of EW.

There are various ways of observing the deviation, the best is by marks at *C* and *D*.

I have not been able to detect any absurdity about this experiment and I think this plan gets rid of all induced currents and makes the two experiments the same every way except the direction of magnetization.

I have just heard of Slessor being Senior Wrangler. I suppose Smith will be 2nd. I got a letter from Hopkins,[b] praising Smith highly. I am glad to learn (from Mr Hardy Robinson) that Mrs Thomson is better this winter.

If you can spare time I should like to hear your opinion on the desirableness of teaching the whole of Statics before the theory of Motion. This year I introduced practical Kinetics in the form of toothed wheels, cranks, Hookes joint &c. which we studied with respect to their *motions* only, bringing in the forces in a different set of lectures. I devoted a little time to more ⟨abstract⟩ theoretical matters such as the motion of the nail of a wheel tracing a cycloid, an ellipse traced by epicycloidal motion &c..

I am beginning to think that the best way is to drive Statics and Kinetics abreast and so prepare for Dynamics.

I intend to lay more carefully the foundations of the theory of absolute and relative 1st position, 2nd velocity, 3rd acceleration. I find that the systematic study of the theory of units and measurement is practically useful and intelligible to the students, and I have had but few bad errors of this kind.

Of course I must teach Statics in the usual way, because it is good training and because that is what men will be examined on afterwards.

By the way what do you think of a board of Examiners, for the Scotch Colleges to ensure a uniform standard of excellence. It would certainly prevent any ambitious deviations from the usual methods on the part of individual professors, for the students would study the examiners text book and not listen to the professor. Now I had a respectable voluntary audience after the hour yesterday to go through the convertible pendulum, and you have experimentalists and essayists. At Cambridge such things are almost crushed by the enlightened Senate House. How much more in Scotland by partially informed examiners.

Yours truly

J. C. Maxwell

a. This letter is misdated 1859. The correct date is 1858; see the penultimate paragraph in the following letter (document 12), dated February 7, 1858. The references to Slessor and Smith show that this letter should precede document 12, together with Maxwell's incomplete description of the ring model given in the first paragraph of this letter.

b. William Hopkins: Maxwell's and Thomson's private tutor at Cambridge, and famous for the number of tripos winners he trained.

12. Letter from Maxwell to R. B. Litchfield, February 7, 1858

Trinity College, Cambridge, Add. MS Letters C 1[89]; quoted in part in Lewis Campbell and William Garnett, *The Life of James Clerk Maxwell* (London: Macmillan and Co., 1882), pp. 301–302.

129 Union Street, Aberdeen
7 Feb. 1858

Dear Litchfield,

It is a long time since I have heard of you, I very much wish to hear how you are and what you are doing or what is doing around you. When I last wrote I was on my way here. Since then I have been at work, Statics and Dynamics; two days a week being devoted to Principles of Mechanism, and afterwards to Friction, Elasticity and Strength of Materials, and also Clocks and Watches, when we come to the pendulum. We have just begun hydrostatics. I have found a better text-book for hydrostatics than I had thought for,—the run of them are so bad, both Cambridge and the other ones,—Galbraith and Haughton's *Manual of Hydrostatics*, Longmans, 2/–. There are also manuals of Mechanics and Optics of the same set. There is no humbug in them, and many practical matters are introduced instead of mere intricacies. The only defect is a somewhat ostentatious resignation of the demonstrations of certain truths, and a leaning upon feigned experiments instead of them. But this is exactly the place where students trust most to the professor, so that I care less about it. I shall adopt the Optics, which have no such defect, and possibly the Mechanics, next year.

My students of last year, to the number of about fourteen, form a voluntary class, and continue their studies. We went through Newton i., ii., iii., and took a rough view of the Lunar Theory, and of the present state of Astronomy. Then we have taken up Magnetism and Electricity, static and current and now we are at Electro-magnetism and Ampère's Laws. I intend to make Faraday's book the backbone of all the rest, as he himself is the nucleus of everything electric since 1830.

So much for class work. Saturn's Rings are going on still, but this month I am clearing out some spare time to work them in. I have got up a model to show the motions of a ring of satellites, a very neat piece of work, by Ramage, the maker of the "top".

For other things—I have not much time in winter for improving my mind. I have read Froude's *History*,[a] *Aurora Leigh*,[b] & Hopkins's *Essay on Geology*,[c] also Herschel's collected *Essays*,[d] which I liked much, also Lavater's life and *Physiognomy*,[e] which has introduced me to him pleasantly though verbosely. I like the man very much, quite apart from his conclusions and dogmas. They are only results, and far inferior to methods. But many of them are true if properly understood and applied, and I suppose the rest are worth respect as statements of a truth-telling man.

Well, work is good, and reading is good, but friends are better. I have but a finite number of friends, and they are dropping off, one here, one there. A few live and flourish. Let it be long, and let us work while there is day, for the night is coming, and work by day leads to rest at night.

How is C. J. Monro? Do you ever hear of Mrs. Pomeroy? What of Farrar?

I suppose you know that Slessor[f] the Senior Wrangler is of Kings College Aberdeen and that Smith the 2nd is of Glasgow. Smith will do something yet, better than 2nd Wrangler. Another Aberdonian (Kings), Stirling, was 1st math. in the May[g] at Trinity.

I think one of my men is inclined that way. I am not encouraging him to it as he is very ambitious, but if he determines on it he will do well, if he keeps out of a small college. He is far too versatile to be trusted where boating, billiards beer &c. are more immediate paths to distinction than the pursuit of wisdom either mathematical, classical or social. If he could, I think he would naturally seek those whom he could respect as his superiors in something or other, and so avoid being the cock of a dunghill.

Yours truly,
James Clerk Maxwell

a. James Anthony Froude (1818–1894) was a disciple and close friend of Carlyle. His frank biography and unexpurgated edition of Carlyle's letters and those of his wife caused a scandal. Maxwell is referring to the first two volumes of Froude's *History of England from the Fall of Wolsey to the Death of Elizabeth* (1856), which established his reputation.

b. Elizabeth Barrett Browning (1806–1861), *Aurora Leigh* (1856).

c. William Hopkins (1793–1866), Maxwell's private tutor at Cambridge, became interested in geology about 1830 and developed mathematical models to explain geological phenomena. The essay to which Maxwell refers is not a text but a long article: Hopkins, "Geology," *Cambridge Essays* (London: J. Parker and Sons, 1855), pp. 172–240. The essay was an introduction to theoretical geology.

d. John William Frederick Herschel (1792–1871), *Essays from the Edinburgh and Quarterly Review* (London: Longman Brown, 1857).

e. Johann Caspar Lavater (1741–1801) was a young, close associate of Goethe. He published his own autobiography, and his correspondence with Goethe was published as *Goethe und Lavater. Briefe und Tagbucher* ed. Heinrich Funck (Weimar: Goethe-gesellschaft, 1901). Maxwell refers to Lavater, *Physiognomische Fragmente zur Beforderung der Menschenkenntnis und Menschenliebe* (Leipzig, Winterthur: Weidmanns Erben und Reich, und Heinreich Steiner, 1775–1778). This gives some idea of the range of Maxwell's intellectual interests.

f. G. M. Slessor: In 1861 he suceeded Tait as professor of natural philosophy at Belfast, when Tait went to Edinburgh.

g. This refers to the college examinations at Trinity.

13. Letter from Maxwell to Lewis Campbell, February 17, 1858

Lewis Campbell and William Garnett, *The Life of James Clerk Maxwell* (London: Macmillan and Co., 1882), pp. 302–303.

... I have not been reading much of late. I have been hard at mathematics. In fact I set myself a great arithmetical job of calculating the tangential action of two rings of satellites, and I am near through with it now. I have got a very neat model of my theoretical ring, a credit to Aberdeen workmen. Here is a diagram, but the thing is complex and difficult to draw:—

Two wheels turning on parallel parts of a cranked axle; thirty-six little cranks of the same length between corresponding points of the circumferences; each carries a little ivory satellite.[1]

[1] The sketch which follows corresponds to the model which is preserved in the Cavendish Laboratory. See Part II.[a]

a. Footnote from Campbell and Garnett. The sketches to which they refer are reproduced on p. 511 of their *Life*. These are also the figures 7 and 8 of Maxwell, "Saturn's Rings," on the page facing p. 285 in the *Scientific Papers*, vol. I (reproduced in document 17).

14. Letter from Maxwell to Cecil J. Munro, April 29, 1858[a]

Lewis Campbell and William Garnett, *The Life of James Clerk Maxwell* (London: Macmillan and Co., 1882), p.309.

Glenlair
29th April, 1858

... I displayed my model of Saturn's Rings at the Edinburgh Royal Society on the 19th. The anatomists seemed to take more interest in the construction of it. We[b] are going to do some experiments on colour this summer, if my prisms turn out well. I have got a beautiful set of slits made by Ramage, to let in the different pencils of light at the proper places, and of the proper breadths.

a. Cecil J. Munro was a college friend of Maxwell.
b. "We" refers to Maxwell and his wife Mary Katherine Dewar, whom Maxwell married in June 1858.

15. Letter from Maxwell to Cecil J. Munro, July 24, 1858

Lewis Campbell and William Garnett, *The Life of James Clerk Maxwell* (London: Macmillan and Co., 1882), p.313.

Glenlair
24th July 1858

... We are no great students at present, preferring various passive enjoyments, resulting from the elemental influences of sun, wind and streams.[a] This week I have begun to make a small hole into Saturn who has slept on his voluminous ring for months.

a. Maxwell's marriage took place on June 2, 1858.

16. Letter from Maxwell to George Gabriel Stokes, September 7, 1858[a]

Memoir and Scientific Correspondence of the Late Sir George Gabriel Stokes, ed. Joseph Larmor 2 vols. (Cambridge: Cambridge University Press, 1907), vol. 2, pp. 7–8.

Glenlair, Springholm, Dumfries
Sept. 7th, 1858

Dear Stokes,

I am just finishing my essay on Saturn's Rings, and there is a problem in it about a set of fluid rings revolving each with its proper velocity with friction against its neighbours. Now I am in want of the *coefficient of internal friction* in water and in air, in order to condescend to numbers. I have your paper on pendulums,[b] but it is locked up at Aberdeen, so I have written to you rather than wait till I go.

It is the coefficient B^c of your paper on elasticity and fluid friction, that is the number of units of force acting tangentially on units of area sliding with unit of relative velocity past a fluid plane at unit of distance, stating units of force and length.

I think you have got it somewhere at hand. I have not.

It is not long since I began to do mathematics after a somewhat long rest. I have set up a neat model for showing the disturbances among a set of satellites forming a ring. It was made by Ramage of Aberdeen, who made my dynamical top.

I have done no optics this summer, but have been vegetating in the country with success.

Yours truly,
James Clerk Maxwell

a. George Gabriel Stokes (1819–1903).

b. The paper to which Maxwell refers is Stokes, "On the Effect of the Internal Friction of Fluids on the Motions of Pendulums," *Transactions of the Cambridge Philosophical Society* 9(1856), 8–106 (read in 1850), abstracted in *Philosophical Magazine* 1(1851), 337–339. It is reprinted in Stokes. *Mathematical and Physical Papers* (Cambridge: Cambridge University Press, 1880–1905), vol. 3, pp. 1–141.

c. In Stokes's notation the coefficient of friction is denoted k or k'.

17

The title page and advertisement are from James Clerk Maxwell, *On the Stability of the Motion of Saturn's Rings* (Cambridge: Macmillan and Co., 1859). The remainder is from *The Scientific Papers of James Clerk Maxwell*, ed. W. D. Niven (New York: Dover, reprint 1965), vol. 1, pp. 289–374 and 2 pages of plates following p. 376.

ON THE

STABILITY OF THE MOTION

OF

SATURN'S RINGS.

AN ESSAY,

WHICH OBTAINED THE ADAMS PRIZE FOR THE YEAR 1856, IN THE
UNIVERSITY OF CAMBRIDGE.

By J. CLERK MAXWELL, M.A.

LATE FELLOW OF TRINITY COLLEGE, CAMBRIDGE.
PROFESSOR OF NATURAL PHILOSOPHY
IN THE MARISCHAL COLLEGE AND UNIVERSITY OF ABERDEEN.

" E pur si muove.'

Cambridge:

MACMILLAN AND CO.

AND 23 HENRIETTA STREET, COVENT GARDEN, LONDON.

1859.

ADVERTISEMENT.

THE Subject of the Prize was announced in the following terms:—

The University having accepted a fund raised by several members of St John's College, for the purpose of founding a Prize to be called the ADAMS PRIZE, for the best Essay on some subject of Pure Mathematics, Astronomy, or other branch of Natural Philosophy, the Prize to be given once in two years, and to be *open to the competition of all persons who have at any time been admitted to a degree in this University:*—

The Examiners give Notice, that the following is the subject for the Prize to be adjudged in 1857:—

The Motions of Saturn's Rings.

*** The Problem may be treated on the supposition that the system of Rings is exactly or very approximately concentric with Saturn and symmetrically disposed about the plane of his Equator, and different hypotheses may be made respecting the physical constitution of the Rings. It may be supposed (1) that they are rigid: (2) that they are fluid, or in part aeriform: (3) that they consist of masses of matter not mutually coherent. The question will be considered to be answered by ascertaining on these hypotheses severally, whether the conditions of mechanical stability are satisfied by the mutual attractions and motions of the Planet and the Rings.

It is desirable that an attempt should also be made to determine on which of the above hypotheses the appearances both of the bright Rings and the recently discovered dark Ring may be most satisfactorily explained; and to indicate any causes to which a change of form, such as is supposed from a comparison of modern with the earlier observations to have taken place, may be attributed.

E. GUEST, *Vice-Chancellor.*

J. CHALLIS.

S. PARKINSON.

W. THOMSON.

March 23, 1855.

CONTENTS.

PART I.

ON THE MOTION OF A RIGID BODY OF ANY FORM ABOUT
A SPHERE.

PART II.

ON THE MOTION OF A RING, THE PARTS OF WHICH ARE NOT
RIGIDLY CONNECTED.

THERE are some questions in Astronomy, to which we are attracted rather on account of their peculiarity, as the possible illustration of some unknown principle, than from any direct advantage which their solution would afford to mankind. The theory of the Moon's inequalities, though in its first stages it presents theorems interesting to all students of mechanics, has been pursued into such intricacies of calculation as can be followed up only by those who make the improvement of the Lunar Tables the object of their lives. The value of the labours of these men is recognised by all who are aware of the importance of such tables in Practical Astronomy and Navigation. The methods by which the results are obtained are admitted to be sound, and we leave to professional astronomers the labour and the merit of developing them.

The questions which are suggested by the appearance of Saturn's Rings cannot, in the present state of Astronomy, call forth so great an amount of labour among mathematicians. I am not aware that any practical use has been made of Saturn's Rings, either in Astronomy or in Navigation. They are too distant, and too insignificant in mass, to produce any appreciable effect on the motion of other parts of the Solar system ; and for this very reason it is diffi-cult to determine those elements of their motion which we obtain so accurately in the case of bodies of greater mechanical importance.

But when we contemplate the Rings from a purely scientific point of view, they become the most remarkable bodies in the heavens, except, perhaps, those still less *useful* bodies—the spiral nebulæ. When we have actually seen that great arch swung over the equator of the planet without any visible connexion, we cannot bring our minds to rest. We cannot simply admit that such is the case, and describe it as one of the observed facts in nature, not admitting or requiring explanation. We must either explain its motion on the principles of mechanics, or admit that, in the Saturnian realms, there can be motion regu-lated by laws which we are unable to explain.

The arrangement of the rings is represented in the figure (1) on a scale of one inch to a hundred thousand miles. S is a section of Saturn through his equator, A, B and C are the three rings. A and B have been known for 200 years. They were mistaken by Galileo for protuberances on the planet itself, or perhaps satellites. Huyghens discovered that what he saw was a thin flat ring not touching the planet, and Ball discovered the division between A and B. Other divisions have been observed splitting these again into concentric rings, but these have not continued visible, the only well-established division being one in the middle of A. The third ring C was first detected by Mr Bond, at Cambridge U.S. on November 15, 1850 ; Mr Dawes, not aware of Mr Bond's discovery, observed it on November 29th, and Mr Lassel a few days later. It gives little light compared with the other rings, and is seen where it crosses the planet as an obscure belt, but it is so transparent that the limb of the planet is visible through it, and this without distortion, shewing that the rays of light have not passed through a transparent substance, but between the scattered particles of a discontinuous stream.

It is difficult to estimate the thickness of the system; according to the best estimates it is not more than 100 miles, the diameter of A being 176,418 miles; so that on the scale of our figure the thickness would be one thousandth of an inch.

Such is the scale on which this magnificent system of concentric rings is constructed; we have next to account for their continued existence, and to reconcile it with the known laws of motion and gravitation, so that by rejecting every hypothesis which leads to conclusions at variance with the facts, we may learn more of the nature of these distant bodies than the telescope can yet ascertain. We must account for the rings remaining suspended above the planet, concentric with Saturn and in his equatoreal plane; for the flattened figure of the section of each ring, for the transparency of the inner ring, and for the gradual approach of the inner edge of the ring to the body of Saturn as deduced from all the recorded observations by M. Otto Struvé (*Sur les dimensions des Anneaux de Saturne*—Recueil de Mémoires Astronomiques, Poulkowa, 15 Nov. 1851). For an account of the general appearance of the rings as seen from the planet, see Lardner on the Uranography of Saturn, *Mem. of the Astronomical Society*, 1853. See also the article "Saturn" in Nichol's *Cyclopædia of the Physical Sciences.*

Our curiosity with respect to these questions is rather stimulated than appeased by the investigations of Laplace. That great mathematician, though occupied with many questions which more imperiously demanded his attention, has devoted several chapters in various parts of his great work, to points connected with the Saturnian System.

He has investigated the law of attraction of a ring of small section on a point very near it (*Méc. Cél.* Liv. III. Chap. VI.), and from this he deduces the equation from which the ratio of the breadth to the thickness of each ring is to be found,

$$e = \frac{R^3}{3a^3}\frac{\rho}{\rho'} = \frac{\lambda(\lambda-1)}{(\lambda+1)(3\lambda^2+1)},$$

where R is the radius of Saturn, and ρ his density; a the radius of the ring, and ρ' its density; and λ the ratio of the breadth of the ring to its thickness. The equation for determining λ when e is given has one negative root which must be rejected, and two roots which are positive while $e < 0.0543$, and impossible when e has a greater value. At the critical value of e, $\lambda = 2.594$ nearly.

The fact that λ is impossible when e is above this value, shews that the ring cannot hold together if the ratio of the density of the planet to that of the ring exceeds a certain value. This value is estimated by Laplace at 1·3, assuming $a = 2R$.

We may easily follow the physical interpretation of this result, if we observe that the forces which act on the ring may be reduced to—

(1) The attraction of Saturn, varying inversely as the square of the distance from his centre.

(2) The centrifugal force of the particles of the ring, acting outwards, and varying directly as the distance from Saturn's polar axis.

(3) The attraction of the ring itself, depending on its form and density, and directed, roughly speaking, towards the centre of its section.

The first of these forces must balance the second somewhere near the mean distance of the ring. Beyond this distance their resultant will be outwards, within this distance it will act inwards.

If the attraction of the ring itself is not sufficient to balance these residual forces, the outer and inner portions of the ring will tend to separate, and the ring will be split up; and it appears from Laplace's result that this will be the case if the density of the ring is less than $\frac{10}{13}$ of that of the planet.

This condition applies to all rings whether broad or narrow, of which the parts are separable, and of which the outer and inner parts revolve with the same angular velocity.

Laplace has also shewn (Liv. v. Chap. iii.), that on account of the oblateness of the figure of Saturn, the planes of the rings will follow that of Saturn's equator through every change of its position due to the disturbing action of other heavenly bodies.

Besides this, he proves most distinctly (Liv. iii. Chap. vi.), that a solid uniform ring cannot possibly revolve about a central body in a permanent manner, for the slightest displacement of the centre of the ring from the centre of the planet would originate a motion which would never be checked, and would inevitably precipitate the ring upon the planet, not necessarily by breaking the ring, but by the inside of the ring falling on the equator of the planet.

He therefore infers that the rings are irregular solids, whose centres of gravity do not coincide with their centres of figure. We may draw the conclusion more formally as follows, "If the rings were solid and uniform, their motion would be unstable, and they would be destroyed. But they are not destroyed, and their motion is stable; therefore they are either not uniform or not solid."

I have not discovered * either in the works of Laplace or in those of more recent mathematicians, any investigation of the motion of a ring either not uniform or not solid. So that in the present state of mechanical science, we do not know whether an irregular solid ring, or a fluid or disconnected ring, can revolve permanently about a central body; and the Saturnian system still remains an unregarded witness in heaven to some necessary, but as yet unknown, development of the laws of the universe.

We know, since it has been demonstrated by Laplace, that a uniform solid ring cannot revolve permanently about a planet. We propose in this Essay to determine the amount and nature of the irregularity which would be required to make a permanent rotation possible. We shall find that the stability of the motion of the ring would be ensured by loading the ring at one point with a

---

* Since this was written, Prof. Challis has pointed out to me three important papers in Gould's *Astronomical Journal :*—Mr G. P. Bond *on the Rings of Saturn* (May 1851) and Prof. B. Pierce of Harvard University *on the Constitution of Saturn's Rings* (June 1851), and *on the Adams' Prize Problem* for 1856 (Sept. 1855). These American mathematicians have both considered the conditions of statical equilibrium of a transverse section of a ring, and have come to the conclusion that the rings, if they move each as a whole, must be very narrow compared with the observed rings, so that in reality there must be a great number of them, each revolving with its own velocity. They have also entered on the question of the fluidity of the rings, and Prof. Pierce has made an investigation as to the permanence of the motion of an irregular solid ring and of a fluid ring. The paper in which these questions are treated at large has not (so far as I am aware) been published, and the references to it in Gould's Journal are intended to give rather a popular account of the results, than an accurate outline of the methods employed. In treating of the attractions of an irregular ring, he makes admirable use of the theory of potentials, but his published investigation of the motion of such a body contains some oversights which are due perhaps rather to the imperfections of popular language than to any thing in the mathematical theory. The only part of the theory of a fluid ring which he has yet given an account of, is that in which he considers the form of the ring at any instant as an ellipse; corresponding to the case where $n = \omega$, and $m = 1$. As I had only a limited time for reading these papers, and as I could not ascertain the methods used in the original investigations, I am unable at present to state how far the results of this essay agree with or differ from those obtained by Prof. Pierce.

heavy satellite about $4\frac{1}{2}$ times the weight of the ring, but this load, besides being inconsistent with the observed appearance of the rings, must be far too artificially adjusted to agree with the natural arrangements observed elsewhere, for a very small error in excess or defect would render the ring again unstable.

We are therefore constrained to abandon the theory of a solid ring, and to consider the case of a ring, the parts of which are not rigidly connected, as in the case of a ring of independent satellites, or a fluid ring.

There is now no danger of the whole ring or any part of it being precipitated on the body of the planet. Every particle of the ring is now to be regarded as a satellite of Saturn, disturbed by the attraction of a ring of satellites at the same mean distance from the planet, each of which however is subject to slight displacements. The mutual action of the parts of the ring will be so small compared with the attraction of the planet, that no part of the ring can ever cease to move round Saturn as a satellite.

But the question now before us is altogether different from that relating to the solid ring. We have now to take account of variations in the form and arrangement of the parts of the ring, as well as its motion as a whole, and we have as yet no security that these variations may not accumulate till the ring entirely loses its original form, and collapses into one or more satellites, circulating round Saturn. In fact such a result is one of the leading doctrines of the "nebular theory" of the formation of planetary systems: and we are familiar with the actual breaking up of fluid rings under the action of "capillary" force, in the beautiful experiments of M. Plateau.

In this essay I have shewn that such a destructive tendency actually exists, but that by the revolution of the ring it is converted into the condition of dynamical stability. As the scientific interest of Saturn's Rings depends at present mainly on this question of their stability, I have considered their motion rather as an illustration of general principles, than as a subject for elaborate calculation, and therefore I have confined myself to those parts of the subject which bear upon the question of the permanence of a given form of motion.

There is a very general and very important problem in Dynamics, the solution of which would contain all the results of this Essay and a great deal more. It is this—

"Having found a particular solution of the equations of motion of any material system, to determine whether a slight disturbance of the motion indicated by the solution would cause a small periodic variation, or a total derangement of the motion."

The question may be made to depend upon the conditions of a maximum or a minimum of a function of many variables, but the theory of the tests for distinguishing maxima from minima by the Calculus of Variations becomes so intricate when applied to functions of several variables, that I think it doubtful whether the physical or the abstract problem will be first solved.

PART I.

ON THE MOTION OF A RIGID BODY OF ANY FORM ABOUT A SPHERE.

W E confine our attention for the present to the motion in the plane of reference, as the interest of our problem belongs to the character of this motion, and not to the librations, if any, from this plane.

Let S (Fig. 2) be the centre of gravity of the sphere, which we may call Saturn, and R that of the rigid body, which we may call the Ring. Join RS, and divide it in G so that

$$SG : GR :: R : S,$$

R and S being the masses of the Ring and Saturn respectively.

Then G will be the centre of gravity of the system, and its position will be unaffected by any mutual action between the parts of the system. Assume G as the point to which the motions of the system are to be referred. Draw GA in a direction fixed in space.

Let $\qquad AGR = \theta$, and $SR = r$,

then $\qquad GR = \dfrac{S}{S+R} r$, and $GS = \dfrac{R}{S+R} r$,

so that the positions of S and R are now determined.

Let BRB' be a straight line through R, *fixed with respect to the substance of the ring*, and let $BRK = \phi$.

This determines the angular position of the ring, so that from the values of r, θ, and ϕ the configuration of the system may be deduced, as far as relates to the plane of reference.

We have next to determine the forces which act between the ring and the sphere, and this we shall do by means of the *potential function* due to the ring, which we shall call V.

The value of V for any point of space S, depends on its position relatively to the ring, and it is found from the equation

$$V = \Sigma \left(\frac{dm}{r'} \right),$$

where dm is an element of the mass of the ring, and r' is the distance of that element from the given point, and the summation is extended over every element of mass belonging to the ring. V will then depend entirely upon the position of the point S relatively to the ring, and may be expressed as a function of r, the distance of S from R, the centre of gravity of the ring, and ϕ, the angle which the line SR makes with the line RB, fixed in the ring.

A particle P, placed at S, will, by the theory of potentials, experience a moving force $P\dfrac{dV}{dr}$ in the direction which tends to increase r, and $P\dfrac{1}{r}\dfrac{dV}{d\phi}$ in a tangential direction, tending to increase ϕ.

Now we know that the attraction of a sphere is the same as that of a particle of equal mass placed at its centre. The forces acting between the sphere and the ring are therefore $S\dfrac{dV}{dr}$ tending to increase r, and a tangential force $S\dfrac{1}{r}\dfrac{dV}{d\phi}$, applied at S tending to increase ϕ. In estimating the effect of this latter force on the ring, we must resolve it into a tangential force $S\dfrac{1}{r}\dfrac{dV}{d\phi}$ acting at R, and a couple $S\dfrac{dV}{d\phi}$ tending to increase ϕ.

We are now able to form the equations of motion for the planet and the ring.

For the planet

$$S \frac{d}{dt} \left\{ \left(\frac{Rr}{S+R} \right)^2 \frac{d\theta}{dt} \right\} = - \frac{R}{S+R} S \frac{dV}{d\phi} \dots\dots\dots\dots (1),$$

$$S \frac{d^2}{dt^2} \left(\frac{Rr}{S+R} \right) - S \frac{Rr}{S+R} \left(\frac{d\theta}{dt} \right)^2 = S \frac{dV}{dr} \dots\dots\dots\dots (2).$$

For the centre of gravity of the ring,

$$R \frac{d}{dt} \left\{ \left(\frac{Sr}{S+R} \right)^2 \frac{d\theta}{dt} \right\} = - \frac{S}{S+R} S \frac{dV}{d\phi} \dots\dots\dots (3),$$

$$R \frac{d^2}{dt^2} \left(\frac{Sr}{S+R} \right) - R \frac{Sr}{S+R} \left(\frac{d\theta}{dt} \right)^2 = S \frac{dV}{dr} \dots\dots\dots (4).$$

For the rotation of the ring about its centre of gravity,

$$Rk^2 \frac{d^2}{dt^2} (\theta + \phi) = S \frac{dV}{d\phi} \dots\dots\dots\dots\dots (5),$$

where k is the radius of gyration of the ring about its centre of gravity.

Equation (3) and (4) are necessarily identical with (1) and (2), and shew that the orbit of the centre of gravity of the ring must be similar to that of the Planet. Equations (1) and (3) are equations of areas, (2) and (4) are those of the radius vector.

Equations (3), (4) and (5) may be thus written,

$$R \left\{ 2r \frac{dr}{dt} \frac{d\theta}{dt} + r^2 \frac{d^2\theta}{dt^2} \right\} + (R+S) \frac{dV}{d\phi} = 0 \dots\dots\dots\dots (6),$$

$$R \left\{ \frac{d^2r}{dt^2} - r \left(\frac{d\theta}{dt} \right)^2 \right\} - (R+S) \frac{dV}{dr} = 0 \dots\dots\dots\dots (7),$$

$$Rk^2 \left(\frac{d^2\theta}{dt^2} + \frac{d^2\phi}{dt^2} \right) - S \frac{dV}{d\phi} = 0 \dots\dots\dots\dots (8).$$

These are the necessary and sufficient data for determining the motion of the ring, the initial circumstances being given.

PROB. I.   To find the conditions under which a uniform motion of the ring is possible.

By a uniform motion is here meant a motion of uniform rotation, during which the position of the centre of the Planet with respect to the ring does not change.

In this case r and ϕ are constant, and therefore V and its differential coefficients are given.   Equation (7) becomes,

$$Rr\left(\frac{d\theta}{dt}\right)^2 + (R+S)\frac{dV}{dr} = 0,$$

which shews that the angular velocity is constant, and that

$$\left(\frac{d\theta}{dt}\right)^2 = -\frac{R+S}{Rr}\frac{dV}{dr} = \omega^2,\ \text{say} \ \dots\dots\dots\dots\dots\dots (9).$$

Hence, $\dfrac{d^2\theta}{dt^2} = 0$, and therefore by equation (8),

$$\frac{dV}{d\phi} = 0 \dots\dots\dots\dots\dots\dots\dots\dots\dots\dots\dots\dots(10).$$

Equations (9) and (10) are the conditions under which the uniform motion is possible, and if they were exactly fulfilled, the uniform motion would go on for ever if not disturbed.   But it does not follow that if these conditions were *nearly* fulfilled, or that if when accurately adjusted, the motion were *slightly* disturbed, the motion would go on for ever *nearly* uniform.   The effect of the disturbance might be either to produce a periodic variation in the elements of the motion, the amplitude of the variation being small, or to produce a displacement which would increase indefinitely, and derange the system altogether. In the one case the motion would be *dynamically stable*, and in the other it would be *dynamically unstable*.   The investigation of these displacements while still very small will form the next subject of inquiry.

Prob. II.   To find the equations of the motion when slightly disturbed.

Let $r = r_0$, $\theta = \omega t$ and $\phi = \phi_0$ in the case of uniform motion, and let

$$r = r_0 + r_1,$$
$$\theta = \omega t + \theta_1,$$
$$\phi = \phi_0 + \phi_1,$$

when the motion is slightly disturbed, where r_1, θ_1, and ϕ_1 are to be treated as small quantities of the first order, and their powers and products are to be neglected.   We may expand $\dfrac{dV}{dr}$ and $\dfrac{dV}{d\phi}$ by Taylor's Theorem,

$$\frac{dV}{dr} = \frac{dV}{dr} + \frac{d^2V}{dr^2}\ r_1 + \frac{d^2V}{dr\,d\phi}\ \phi_1,$$

$$\frac{dV}{d\phi} = \frac{dV}{d\phi} + \frac{d^2V}{dr\,d\phi}\ r_1 + \frac{d^2V}{d\phi^2}\ \phi_1,$$

where the values of the differential coefficients on the right-hand side of the equations are those in which r_0 stands for r, and ϕ_0 for ϕ.

Calling
$$\frac{d^2V}{dr^2} = L, \quad \frac{d^2V}{drd\phi} = M, \quad \frac{d^2V}{d\phi^2} = N,$$

and taking account of equations (9) and (10), we may write these equations,

$$\frac{dV}{dr} = -\frac{Rr_0}{R+S}\omega^2 + Lr_1 + M\phi_1,$$

$$\frac{dV}{d\phi} = Mr_1 + N\phi_1.$$

Substituting these values in equations (6), (7), (8), and retaining all small quantities of the first order while omitting their powers and products, we have the following system of linear equations in r_1, θ_1, and ϕ_1,

$$R\left(2r_0\omega\frac{dr_1}{dt} + r_0^2\frac{d^2\theta_1}{dt^2}\right) + (R+S)(Mr_1 + N\phi_1) = 0 \dots\dots(11),$$

$$R\left(\frac{d^2r_1}{dt^2} - \omega^2 r_1 - 2r_0\omega\frac{d\theta_1}{dt}\right) - (R+S)(Lr_1 + M\phi_1) = 0 \dots\dots(12),$$

$$Rk^2\left(\frac{d^2\theta_1}{dt^2} + \frac{d\phi_1}{dt^2}\right) - S(Mr_1 + N\phi_1) = 0 \dots\dots(13).$$

PROB. III. To reduce the three simultaneous equations of motion to the form of a single linear equation.

Let us write n instead of the symbol $\frac{d}{dt}$, then arranging the equations in terms of r_1, θ_1, and ϕ_1, they may be written:

$$\{2R_0\omega n + (R+S)M\}r_1 + (Rr_0^2n^2)\theta_1 + (R+S)N\phi_1 = 0 \dots\dots(14),$$

$$\{Rn^2 - R\omega^2 - (R+S)L\}r_1 - (2Rr_0\omega n)\theta_1 - (R+S)M\phi_1 = 0 \dots\dots(15),$$

$$-(SM)r_1 + (Rk^2n^2)\theta_1 + (Rk^2n^2 - SN)\phi_1 = 0 \dots\dots(16).$$

Here we have three equations to determine three quantities r_1, θ_1, ϕ_1; but it is evident that only a relation can be determined between them, and that in the process for finding their absolute values, the three quantities will vanish together, and leave the following relation among the coefficients,

$$\left.\begin{array}{l} -\{2Rr_0\omega n + (R+S)\,M\}\,\{2Rr_0\omega n\}\,\{Rk^2n^2 - SN\} \\ +\{Rn^2 - R\omega^2 - (R+S)\,L\}\,\{Rk^2n^2\}\,\{(R+S)\,N\} \\ +(SM)\,(Rr_0^2n^2)\,(R+S)\,M - (SM)\,(2Rr_0\omega n)\,(R+S)\,N \\ +\{2Rr_0\omega n + (R+S)\,M\}\,\{Rk^2n^2\}\,\{(R+S)\,M\} \\ -\{Rn^2 - R\omega^2 - (R+S)\}\,\{Rr_0^2n^2\}\,\{Rk^2n^2 - SN\} \end{array}\right\} = 0 \ \ldots\ldots(17).$$

By multiplying up, and arranging by powers of n and dividing by Rn^2, this equation becomes

$$An^4 + Bn^2 + C = 0 \ \ldots\ldots\ldots\ldots\ldots\ldots\ldots\ldots\ldots (18),$$

where

$$\left.\begin{array}{l} A = R^2r_0^2k^2, \\ B = 3R^2r_0^2k^2\omega^2 - R\,(R+S)\,Lr_0^2k^2 - R\,\{(R+S)\,k^2 + Sr^2\}\,N \\ C = R\,\{(R+S)\,k^2 - 3Sr_0^2\}\,\omega^2 + (R+S)\,\{(R+S)\,k^2 + Sr_0^2\}\,(LN - M^2) \end{array}\right\} \ \ldots\ldots(19).$$

Here we have a biquadratic equation in n which may be treated as a quadratic in n^2, it being remembered that n stands for the operation $\dfrac{d}{dt}$.

Prob. IV. To determine whether the motion of the ring is stable or unstable, by means of the relations of the coefficients A, B, C.

The equations to determine the forms of r_1, θ_1, and ϕ_1 are all of the form

$$A\,\frac{d^4u}{dt^4} + B\,\frac{d^2u}{dt^2} + Cu = 0 \ \ldots\ldots\ldots\ldots\ldots\ldots\ldots (20),$$

and if n be one of the four roots of equation (18), then

$$u = De^{nt}$$

will be one of the four terms of the solution, and the values of r_1, θ_1, and ϕ_1 will differ only in the values of the coefficient D.

Let us inquire into the nature of the solution in different cases.

(1) If n be positive, this term would indicate a displacement which must increase indefinitely, so as to destroy the arrangement of the system.

(2) If n be negative, the disturbance which it belongs to would gradually die away.

(3) If n be a pure impossible quantity, of the form $\pm a\sqrt{-1}$, then there will be a term in the solution of the form $D\cos(at+a)$, and this would indicate a periodic variation, whose amplitude is D, and period $\dfrac{2\pi}{a}$.

(4) If n be of the form $b \pm \sqrt{-1}a$, the first term being positive and the second impossible, there will be a term in the solution of the form

$$D\epsilon^{bt} \cos{(at + a)},$$

which indicates a periodic disturbance, whose amplitude continually increases till it disarranges the system.

(5) If n be of the form $-b \pm \sqrt{-1}a$, a negative quantity and an impossible one, the corresponding term of the solution is

$$D\epsilon^{-bt} \cos{(at + a)},$$

which indicates a periodic disturbance whose amplitude is constantly diminishing.

It is manifest that the first and fourth cases are inconsistent with the permanent motion of the system. Now since equation (18) contains only even powers of n, it must have pairs of equal and opposite roots, so that every root coming under the second or fifth cases, implies the existence of another root belonging to the first or fourth. If such a root exists, some disturbance may occur to produce the kind of derangement corresponding to it, so that the system is not safe unless roots of the first and fourth kinds are altogether excluded. This cannot be done without excluding those of the second and fifth kinds, so that, to insure stability, all the four roots must be of the third kind, that is, pure impossible quantities.

That this may be the case, both values of n^2 must be real and negative, and the conditions of this are—

1st.   That $A,$ $B,$ and C should be of the same sign,

2ndly.   That $B^2 > 4AC.$

When these conditions are fulfilled, the disturbances will be periodic and consistent with stability. When they are not both fulfilled, a small disturbance may produce total derangement of the system.

Prob. V.   To find the centre of gravity, the radius of gyration, and the variations of the potential near the centre of a circular ring of small but variable section.

Let a be the radius of the ring, and let θ be the angle subtended at the centre between the radius through the centre of gravity and the line through a given point in the ring. Then if μ be the mass of unit of length of the ring near the given point, μ will be a periodic function of θ, and may there-fore be expanded by Fourier's theorem in the series,

$$\mu = \frac{R}{2\pi a}\{1 + 2f\cos\theta + \tfrac{2}{3}g\cos 2\theta + \tfrac{2}{3}h\sin 2\theta + 2i\cos(3\theta + a) + \&c.\}\ldots\ldots(21),$$

where f, g, h, &c. are arbitrary coefficients, and R is the mass of the ring.

(1) The moment of the ring about the diameter perpendicular to the prime radius is

$$Rr_0 = \int_0^{2\pi} \mu a^2 \cos\theta d\theta = Raf,$$

therefore the distance of the centre of gravity from the centre of the ring,

$$r_0 = af.$$

(2) The radius of gyration of the ring about its centre in its own plane is evidently the radius of the ring $= a$, but if k be that about the centre of gravity, we have

$$k^2 + r_0^2 = a^2\,;$$
$$\therefore\ k^2 = a^2(1 - f^2).$$

(3) The potential at any point is found by dividing the mass of each element by its distance from the given point, and integrating over the whole mass.

Let the given point be near the centre of the ring, and let its position be defined by the co-ordinates r' and ψ, of which r' is small compared with a.

The distance (ρ) between this point and a point in the ring is

$$\frac{1}{\rho} = \frac{1}{a}\left\{1 + \frac{r'}{a}\cos(\psi - \theta) + \tfrac{1}{4}\left(\frac{r'}{a}\right)^2 + \tfrac{3}{4}\left(\frac{r'}{a}\right)^2\cos 2(\psi - \theta) + \&c.\right\}.$$

The other terms contain powers of $\dfrac{r'}{a}$ higher than the second.

We have now to determine the value of the integral,

$$V = \int_0^{2\pi} \frac{\mu}{\rho} ad\theta\,;$$

and in multiplying the terms of (μ) by those of $\left(\dfrac{1}{\rho}\right)$, we need retain only those which contain constant quantities, for all those which contain sines or cosines of multiples of $(\psi - \theta)$ will vanish when integrated between the limits. In this way we find

$$V = \frac{R}{a} \left\{ 1 + f \frac{r'}{a} \cos \psi + \tfrac{1}{4} \frac{r'^2}{a^2} (1 + g \cos 2\psi + h \sin 2\psi) \right\} \dots\dots\dots (22).$$

The other terms containing higher powers of $\dfrac{r'}{a}$.

In order to express V in terms of r_1 and ϕ_1, as we have assumed in the former investigation, we must put

$$r' \cos \psi = -r_1 + \tfrac{1}{2} r_0 \phi_1{}^2,$$
$$r' \sin \psi = -r_0 \phi_1,$$

$$V = \frac{R}{a} \left\{ 1 - f \frac{r_1}{a} + \tfrac{1}{4} \frac{r_1{}^2}{a^2} (1 + g) + \tfrac{1}{2} \frac{h}{a} f r_1 \phi_1 + \tfrac{1}{4} f^2 \phi_1{}^2 (3 - g) \right\} \dots\dots\dots (23).$$

From which we find $\left(\dfrac{dV}{dr} \right)_0 = -\dfrac{R}{a^2} f,$

$$\left. \begin{aligned}
\left(\frac{d^2 V}{dr^2} \right)_0 &= L = \frac{R}{2a^3} (1 + g) \\[4pt]
\left(\frac{d^2 V}{dr\,d\phi} \right)_0 &= M = \frac{R}{2a^2} fh \\[4pt]
\left(\frac{d^2 V}{d\phi^2} \right)_0 &= N = \frac{R}{2a} f^2 (3 - g)
\end{aligned} \right\} \dots\dots\dots\dots\dots\dots (24).$$

These results may be confirmed by the following considerations applicable to any circular ring, and not involving any expansion or integration. Let af be the distance of the centre of gravity from the centre of the ring, and let the ring revolve about its centre with velocity ω. Then the force necessary to keep the ring in that orbit will be $-Raf\omega^2$.

But let S be a mass fixed at the centre of the ring, then if

$$\omega^2 = \frac{S}{a^3},$$

every portion of the ring will be separately retained in its orbit by the attraction of S, so that the whole ring will be retained in its orbit. The resultant attraction must therefore pass through the centre of gravity, and be

$$-Raf\omega^2 = -RS \frac{f}{a^2} ;$$

therefore

$$\frac{dV}{dr} = -R \frac{f}{a^2}.$$

The equation $$\frac{d^2V}{dx^2} + \frac{d^2V}{dy^2} + \frac{d^2V}{dz^2} + 4\pi\rho = 0$$

is true for any system of matter attracting according to the law of gravitation. If we bear in mind that the expression is identical in form with that which measures the total efflux of fluid from a differential element of volume, where $\frac{dV}{dx}, \frac{dV}{dy}, \frac{dV}{dz}$ are the rates at which the fluid passes through its sides, we may easily form the equation for any other case. Now let the position of a point in space be determined by the co-ordinates r, ϕ and z, where z is measured perpendicularly to the plane of the angle ϕ. Then by choosing the directions of the axes x, y, z, so as to coincide with those of the radius vector r, the perpendicular to it in the plane of ϕ, and the normal, we shall have

$$dx = dr, \qquad dy = r\,d\phi, \qquad dz = dz,$$

$$\frac{dV}{dx} = \frac{dV}{dr}, \quad \frac{dV}{dy} = \frac{1}{r}\frac{dV}{d\phi}, \quad \frac{dV}{dz} = \frac{dV}{dz}.$$

The quantities of fluid passing through an element of area in each direction are

$$\frac{dV}{dr}\,r\,d\phi\,dz, \quad \frac{dV}{d\phi}\frac{1}{r}\,dr\,dz, \quad \frac{dV}{dz}\,r\,d\phi\,dr,$$

so that the expression for the whole efflux is

$$\frac{1}{r}\frac{dV}{dr} + \frac{d^2V}{dr^2} + \frac{1}{r^2}\frac{d^2V}{d\phi^2} + \frac{d^2V}{dz^2} \quad\ldots\ldots\ldots\ldots\ldots\ldots (25),$$

which is necessarily equivalent to the former expression.

Now at the centre of the ring $\frac{d^2V}{dz^2}$ may be found by considering the attraction on a point just above the centre at a distance z,

$$\frac{dV}{dz} = -R\frac{z}{(a^2+z^2)^{\frac{3}{2}}},$$

$$\frac{d^2V}{dz^2} = -\frac{R}{a^3}, \text{ when } z=0.$$

Also we know $\qquad \frac{1}{r}\frac{dV}{dr} = -\frac{R}{a^3}$, and $r = af$,

so that in any circular ring $\qquad \frac{d^2V}{dr^2} + \frac{1}{a^2f^2}\frac{d^2V}{d\phi^2} = 2\frac{R}{a^3} \quad\ldots\ldots\ldots\ldots\ldots\ldots (26),$

an equation satisfied by the former values of L and N.

By referring to the original expression for the variable section of the ring, it appears that the effect of the coefficient f is to make the ring thicker on one side and thinner on the other in a uniformly graduated manner. The effect of g is to thicken the ring at two opposite sides, and diminish its section in the parts between. The coefficient h indicates an inequality of the same kind, only not symmetrically disposed about the diameter through the centre of gravity.

Other terms indicating inequalities recurring three or more times in the circumference of the ring, have no effect on the values of L, M and N. There is one remarkable case, however, in which the irregularity consists of a single heavy particle placed at a point on the circumference of the ring.

Let P be the mass of the particle, and Q that of the uniform ring on which it is fixed, then $R = P + Q$,

$$f = \frac{P}{R},$$

$$L = 2\,\frac{P}{a^3} + \frac{Q}{2a^3} = \frac{P+Q}{2a^3}\left(1 + 3\,\frac{P}{R}\right) = \frac{R}{2a^3}(1+g)\,;$$

$$\therefore\; g = \frac{3P}{R} = 3f \dots\dots\dots\dots\dots\dots\dots\dots(27).$$

PROB. VI. To determine the conditions of stability of the motion in terms of the coefficients f, g, h, which indicate the distribution of mass in the ring.

The quantities which enter into the differential equation of motion (18) are R, S, k^2, r_0, ω^2, L, M, N. We must observe that S is very large compared with R, and therefore we neglect R in those terms in which it is added to S, and we put

$$S = a^3\omega^2,$$

$$k^2 = a^2\left(1 - f^2\right),$$

$$r_0 = af,$$

$$L = \frac{R}{2a^3}\left(1 + g\right),$$

$$M = \frac{R}{2a^2}fh,$$

$$N = \frac{R}{2a}\,f^2\left(3 - g\right).$$

Substituting these values in equation (18) and dividing by $R^2a'f^2$, we obtain

$$(1-f^2)\, n^4 + (1 - \tfrac{5}{2}f^2 + \tfrac{1}{2}f^2g)\, n^2\omega^2 + (\tfrac{9}{4} - 6f^2 - \tfrac{1}{4}g^2 - \tfrac{1}{4}h^2 + 2f^2g)\, \omega^4 = 0 \ldots\ldots(28).$$

The condition of stability is that this equation shall give both values of n^2 negative, and this renders it necessary that all the coefficients should have the same sign, and that the square of the second should exceed four times the product of the first and third.

(1) Now if we suppose the ring to be uniform, f, g and h disappear, and the equation becomes

$$n^4 + n^2\omega^2 + \tfrac{9}{4} = 0 \ldots\ldots\ldots\ldots\ldots\ldots\ldots\ldots\ldots (29),$$

which gives impossible values to n^2 and indicates the instability of a uniform ring.

(2) If we make g and $h=0$, we have the case of a ring thicker at one side than the other, and varying in section according to the simple law of sines. We must remember, however, that f must be less than $\tfrac{1}{2}$, in order that the section of the ring at the thinnest part may be real. The equation becomes

$$(1-f^2)\, n^4 + (1 - \tfrac{5}{2}f^2)\, n^2\omega^2 + (\tfrac{9}{4} - 6f^2)\, \omega^4 = 0 \ldots\ldots\ldots\ldots(30).$$

The condition that the third term should be positive gives

$$f^2 < {\cdot}375.$$

The condition that n^2 should be real gives

$$71f^4 - 112f^2 + 32 \text{ negative,}$$

which requires f^2 to be between $\cdot 37445$ and $1\cdot 2$.

The condition of stability is therefore that f^2 should lie between

$$\cdot 37445 \text{ and } \cdot 375,$$

but the construction of the ring on this principle requires that f^2 should be less than $\cdot 25$, so that it is impossible to reconcile this form of the ring with the conditions of stability.

(3) Let us next take the case of a uniform ring, loaded with a heavy particle at a point of its circumference. We have then $g = 3f$, $h = 0$, and the equation becomes

$$(1-f^2)\, n^4 + (1 - \tfrac{5}{2}f^2 + \tfrac{3}{2}f^3)\, n^2\omega^2 + (\tfrac{9}{4} - \tfrac{33}{4}f^2 + 6f^3)\, \omega^4 = 0 \ldots\ldots\ldots\ldots (31).$$

Dividing each term by $1-f$, we get

$$(1+f)\, n^4 + (1+f - \tfrac{3}{2}f^2)\, n^2\omega^2 + \tfrac{3}{4}\{3(1+f) - 8f^2\}\, \omega^4 = 0 \dots\dots\dots\dots(32).$$

The first condition gives f less than $\cdot8279$.

The second condition gives f greater than $\cdot815865$.

Let us assume as a particular case between these limits $f = \cdot82$, which makes the ratio of the mass of the particle to that of the ring as 82 to 18, then the equation becomes

$$1\cdot82\, n^4 + \cdot8114\, n^2\omega^2 + \cdot9696\omega^4 = 0 \dots\dots\dots\dots\dots (33),$$

which gives $\qquad\qquad \sqrt{-1}\,n = \pm\cdot5916\omega \text{ or } \pm\cdot3076\omega.$

These values of n indicate variations of r_1, θ_1, and ϕ_1, which are compounded of two simple periodic inequalities, the period of the one being $1\cdot69$ revolutions, and that of the other $3\cdot251$ revolutions of the ring. The relations between the phases and amplitudes of these inequalities must be deduced from equations (14), (15), (16), in order that the character of the motion may be completely determined.

Equations (14), (15), (16) may be written as follows:

$$(4n\omega + h\omega^2)\,\frac{r_1}{a} + 2fn^2\theta_1 + f(3-g)\,\omega^2\phi_1 = 0 \dots\dots\dots\dots (34),$$

$$\{n^2 - \tfrac{1}{2}\omega^2(3+g)\}\,\frac{r_1}{a} - 2f\omega n\theta_1 - \tfrac{1}{2}fh\omega^2\phi_1 = 0 \dots\dots\dots\dots(35),$$

$$-fh\omega^2\,\frac{r_1}{a} + 2(1-f^2)\,n^2\theta_1 + \{2(1-f^2)\,n^2 - f^2(3-g)\,\omega^2\}\,\phi_1 = 0 \dots\dots (36).$$

By eliminating one of the variables between any two of these equations, we may determine the relation between the two remaining variables. Assuming one of these to be a periodic function of t of the form $A\cos\nu t$, and remembering that n stands for the operation $\dfrac{d}{dt}$, we may find the form of the other.

Thus, eliminating θ_1 between the first and second equations,

$$\{n^3 + \tfrac{1}{2}n\omega^2(5-g) + h\omega^3\}\frac{r_1}{a} + f\omega^2\{(3-g)\,\omega - \tfrac{1}{2}hn\}\,\phi_1 = 0 \dots\dots\dots\dots (37).$$

Assuming $\qquad \dfrac{r_1}{a} = A \sin \nu t$, and $\phi_1 = Q \cos (\nu t - \beta)$,

$$\{-\nu^3 + \tfrac{1}{2}\nu\omega^2 (5-g)\} A \cos \nu t + h\omega^3 A \sin \nu t + f\omega^3 (3-g) Q \cos (\nu t - \beta) + \tfrac{1}{2} f h\omega^2 \nu Q \sin (\nu t - \beta).$$

Equating νt to 0, and to $\dfrac{\pi}{2}$, we get the equations

$$\{\nu^3 - \tfrac{1}{2}\nu\omega^2 (5-g)\} A = f\omega^2 Q \{(3-g)\, \omega \cos \beta - \tfrac{1}{2}h\nu \sin \beta\},$$

$$-h\omega^2 A = f\omega^2 Q \{(3-g)\, \omega \sin \beta + \tfrac{1}{2}h\nu \cos \beta\},$$

from which to determine Q and β.

In all cases in which the mass is disposed symmetrically about the diameter through the centre of gravity, $h = 0$ and the equations may be greatly simplified.

Let $\theta_1 = P \cos (\nu t - a)$, then the second equation becomes

$$\{\nu^2 + \tfrac{1}{2}\omega^2 (3+g)\} A \sin \nu t = 2Pf\omega\nu \sin (\nu t - a),$$

whence $\qquad\qquad a = 0, \quad P = \dfrac{\nu^2 + \tfrac{1}{2}\omega^2 (3+g)}{2f\omega\nu} A \quad\dotfill (38).$

The first equation becomes

$$4A\omega\nu \cos \nu t - 2Pf\nu^2 \cos \nu t + Qf (3-g)\, \omega^2 \cos (\nu t - \beta) = 0,$$

whence $\qquad\qquad \beta = 0, \quad Q = \dfrac{\nu^3 - \tfrac{1}{2}\,\omega^2\nu\,(5-g)}{f(3-g)\,\omega^3} A \quad\dotfill (39).$

In the numerical example in which a heavy particle was fixed to the circumference of the ring, we have, when $f = {\cdot}82$,

$$\frac{\nu}{\omega} = \begin{cases} {\cdot}5916 \\ {\cdot}3076 \end{cases}, \quad \frac{P}{A} = \begin{cases} 3{\cdot}21 \\ 5{\cdot}72 \end{cases}, \quad \frac{Q}{A} = \begin{cases} -1{\cdot}229 \\ -\ {\cdot}797 \end{cases},$$

so that if we put $\omega t = \theta_0 =$ the mean anomaly,

$$\frac{r_1}{a} = A \sin ({\cdot}5916\, \theta_0 - a) + B \sin ({\cdot}3076\, \theta_0 - \beta) \quad\dotfill (40),$$

$$\theta_1 = 3{\cdot}21 A \cos ({\cdot}5916\, \theta_0 - a) + 5{\cdot}72 B \cos ({\cdot}3076\, \theta_0 - \beta) \dotfill (41),$$

$$\phi_1 = -1{\cdot}229 A \cos ({\cdot}5916\, \theta_0 - a) - 5{\cdot}797 B \cos ({\cdot}3076\, \theta_0 - \beta) \ \dots (42).$$

These three equations serve to determine r_1, θ_1 and ϕ_1 when the original motion is given. They contain four arbitrary constants A, B, a, β. Now since the original values r_1, θ_1, ϕ_1, and also their first differential coefficients with respect to t, are arbitrary, it would appear that six arbitrary constants ought to enter into the equation. The reason why they do not is that we assume r_0 and θ_0 as the *mean values* of r and θ in the *actual motion*. These quantities therefore depend on the original circumstances, and the two additional arbitrary constants enter into the values of r_0 and θ_0. In the analytical treatment of the problem the differential equation in n was originally of the sixth degree with a solution $n^2 = 0$, which implies the possibility of terms in the solution of the form $Ct + D$.

The existence of such terms depends on the previous equations, and we find that a term of this form may enter into the value of θ, and that r_1 may contain a constant term, but that in both cases these additions will be absorbed into the values of θ_0 and r_0.

PART II.

ON THE MOTION OF A RING, THE PARTS OF WHICH ARE NOT RIGIDLY CONNECTED.

1. In the case of the Ring of invariable form, we took advantage of the principle that the mutual actions of the parts of any system form at all times a system of forces in equilibrium, and we took no account of the attraction between one part of the ring and any other part, since no motion could result from this kind of action. But when we regard the different parts of the ring as capable of independent motion, we must take account of the attraction on each portion of the ring as affected by the irregularities of the other parts, and therefore we must begin by investigating the statical part of the problem in order to determine the forces that act on any portion of the ring, as depending on the instantaneous condition of the rest of the ring.

In order to bring the problem within the reach of our mathematical methods, we limit it to the case in which the ring is *nearly* circular and uniform, and has a transverse section very small compared with the radius of the ring. By analysing the difficulties of the theory of a linear ring, we shall be better able to appreciate those which occur in the theory of the actual rings.

The ring which we consider is therefore small in section, and very nearly circular and uniform, and revolving with nearly uniform velocity. The variations from circular form, uniform section, and uniform velocity must be expressed by a proper notation.

2.   To express the position of an element of a variable ring at a given time in terms of the original position of the element in the ring.

Let S (fig. 3) be the central body, and SA a direction fixed in space.

Let SB be a radius, revolving with the mean angular velocity ω of the ring, so that $ASB = \omega t$.

Let π be an element of the ring in its actual position, and let P be the position it would have had if it had moved uniformly with the mean velocity ω and had not been displaced, then BSP is a constant angle $= s$, and the value of s enables us to identify any element of the ring.

The element may be removed from its mean position P in three different ways.

(1)   By change of distance from S by a quantity $p\pi = \rho$.

(2)   By change of angular position through a space $Pp = \sigma$.

(3)   By displacement perpendicular to the plane of the paper by a quantity ζ.

ρ, σ and ζ are all functions of s and t. If we could calculate the attractions on any element as depending on the form of these functions, we might determine the motion of the ring for any given original disturbance. We cannot, however, make any calculations of this kind without knowing the form of the functions, and therefore we must adopt the following method of separating the original disturbance into others of simpler form, first given in Fourier's *Traité de Chaleur*.

3.   Let U be a function of s, it is required to express U in a series of sines and cosines of multiples of s between the values $s = 0$ and $s = 2\pi$.

Assume $U = A_1 \cos s + A_2 \cos 2s + \&c. + A_m \cos ms + A_n \cos ns$

$+ B_1 \sin s + B_2 \cos 2s + \&c. + B_m \sin ms + B_n \sin ns.$

Multiply by cos $msds$ and integrate, then all terms of the form

$$\int \cos ms \cos nsds \text{ and } \int \cos ms \sin nsds$$

will vanish, if we integrate from $s=0$ to $s=2\pi$, and there remains

$$\int_0^{2\pi} U \cos msds = \pi A_m, \quad \int_0^{2\pi} U \sin msds = \pi B_m.$$

If we can determine the values of these integrals in the given case, we can find the proper coefficients A_m, B_m, &c., and the series will then represent the values of U from $s=0$ to $s=2\pi$, whether those values be continuous or discontinuous, and when none of those values are infinite the series will be convergent.

In this way we may separate the most complex disturbances of a ring into parts whose form is that of a circular function of s or its multiples. Each of these partial disturbances may be investigated separately, and its effect on the attractions of the ring ascertained either accurately or approximately.

4. To find the magnitude and direction of the attraction between two elements of a disturbed ring.

Let P and Q (fig. 4) be the two elements, and let their original positions be denoted by s_1 and s_2, the values of the arcs BP, BQ before displacement. The displacement consists in the angle BSP being increased by σ_1 and BSQ by σ_2, while the distance of P from the centre is increased by ρ_1 and that of Q by ρ_2. We have to determine the effect of these displacements on the distance PQ and the angle SPQ.

Let the radius of the ring be unity, and $s_2 - s_1 = 2\theta$, then the original value of PQ will be $2 \sin \theta$, and the increase due to displacement

$$= (\rho_2 + \rho_1) \sin \theta + (\sigma_2 - \sigma_1) \cos \theta.$$

We may write the complete value of PQ thus,

$$PQ = 2 \sin \theta \{1 + \tfrac{1}{2} (\rho_2 + \rho_1) + \tfrac{1}{2} (\sigma_2 - \sigma_1) \cot \theta\} \dots\dots\dots\dots (1).$$

The original value of the angle SPQ was $\dfrac{\pi}{2} - \theta$, and the increase due to

displacement is $\quad\quad\quad\quad \tfrac{1}{2} (\rho_2 - \rho_1) \cot \theta - \tfrac{1}{2} (\sigma_2 - \sigma_1),$

so that we may write the values of $\sin SPQ$ and $\cos SPQ$,

$$\sin SPQ = \cos\theta\,\{1+\tfrac{1}{2}(\rho_2-\rho_1)-\tfrac{1}{2}(\sigma_2-\sigma_1)\tan\theta\}\quad\ldots\ldots\ldots\ldots (2),$$

$$\cos SPQ = \sin\theta\,\{1-\tfrac{1}{2}(\rho_2-\rho_1)\cot^2\theta+\tfrac{1}{2}(\sigma_2-\sigma_1)\cot\theta\}\quad\ldots\ldots (3).$$

If we assume the masses of P and Q each equal to $\dfrac{1}{\mu}R$, where R is the mass of the ring, and μ the number of satellites of which it is composed, the accelerating effect of the radial force on P is

$$\frac{1}{\mu}R\frac{\cos SPQ}{PQ^2}=\frac{1}{\mu}\frac{R}{4\sin\theta}\{1-(\rho_2+\rho_1)-\tfrac{1}{2}(\rho_2-\rho_1)\cot^2\theta-\tfrac{1}{2}(\sigma_2-\sigma_1)\cot\theta\}\ldots (4),$$

and the tangential force

$$\frac{1}{\mu}R\frac{\sin SPQ}{PQ^2}=\frac{1}{\mu}\frac{R\cos\theta}{4\sin^2\theta}\{1-\tfrac{1}{2}\rho_2-\tfrac{3}{2}\rho_1-(\sigma_2-\sigma_1)(\cot\theta+\tfrac{1}{2}\tan\theta)\}\ldots\ldots (5).$$

The normal force is $\dfrac{1}{\mu}R\dfrac{\zeta_2-\zeta_1}{8\sin^3\theta}$.

5. Let us substitute for ρ, σ and ζ their values expressed in a series of sines and cosines of multiples of s, the terms involving ms being

$$\rho_1 = A\cos(ms+a), \qquad \rho_2 = A\cos(ms+a+2\theta),$$
$$\sigma_1 = B\sin(ms+\beta), \qquad \sigma_2 = B\sin(ms+\beta+2\theta),$$
$$\zeta_1 = C\cos(ms+\gamma), \qquad \zeta_2 = C\cos(ms+\gamma+2\theta).$$

The radial force now becomes

$$\frac{1}{\mu}\frac{R}{4\sin\theta}\left\{\begin{array}{l}1-A\cos(ms+a)(1+\cos 2m\theta)+A\sin(ms+a)\sin 2m\theta\\[4pt]+\tfrac{1}{2}A\cos(ms+a)(1-\cos 2m\theta)\cot^2\theta-\tfrac{1}{2}A\sin(ms+a)\sin 2m\theta\cot^2\theta\\[4pt]+\tfrac{1}{2}B\sin(ms+\beta)(1-\cos 2m\theta)\cot\theta-\tfrac{1}{2}B\cos(ms+\beta)\sin 2m\theta\cot\theta\end{array}\right\}\ (6).$$

The radial component of the attraction of a corresponding particle on the other side of P may be found by changing the sign of θ. Adding the two together, we have for the effect of the pair

$$\frac{1}{\mu}\frac{R}{2\sin\theta}\{1-A\cos(ms+a)(2\cos^2 m\theta-\sin^2 m\theta\cot^2\theta)$$

$$-B\cos(ms+\beta)\tfrac{1}{2}\sin 2m\theta\cot\theta\}\quad\ldots\ldots\ldots\ldots (7).$$

Let us put

$$L = \Sigma \left(\tfrac{1}{2} \frac{\sin^2 m\theta \cos^2 \theta}{\sin^3 \theta} - \frac{\cos^2 m\theta}{\sin \theta} \right)$$

$$M = \Sigma \left(\frac{\sin 2m\theta \cos \theta}{4 \sin^2 \theta} \right)$$

$$N = \Sigma \left(\frac{\sin^2 m\theta \cos^2 \theta}{\sin^3 \theta} + \tfrac{1}{2} \frac{\sin^2 m\theta}{\sin \theta} \right) \Bigg\} \quad\dots\dots\dots\dots (8)^{*},$$

$$J = \Sigma \left(\frac{\sin^2 m\theta}{2 \sin^3 \theta} \right)$$

$$K = \Sigma \left(\frac{1}{2 \sin \theta} \right)$$

where the summation extends to all the satellites on the same side of P, that is, every value of θ of the form $\frac{x}{\mu} \pi$, where x is a whole number less than $\frac{\mu}{2}$.

The radial force may now be written

$$P = \frac{1}{\mu} R \{ K + LA \cos (ms + a) - MB \cos (ms + \beta) \} \quad\dots\dots\dots\dots (9).$$

---

* The following values of several quantities which enter into these investigations are calculated for a ring of 36 satellites.

$$K = 24 \cdot 5.$$

	$\Sigma \frac{\sin^2 m\theta \cos^2 \theta}{\sin^3 \theta}$	$\Sigma \frac{\cos^2 m\theta}{\sin \theta}$	L	M	N
$m = 0$	0	43	−43	0	0
$m = 1$	32	32	−16	16	37
$m = 2$	107	28	26	25	115
$m = 3$	212	25	81	28	221
$m = 4$	401	24	177	32	411
$m = 9$	975	20	468	30	986
$m = 18$	1569	18	767	0	1582

When μ is very great,

$$\left.\frac{\pi}{\mu}\right|^3 L = \cdot 5259 \text{ when } m = \frac{\mu}{2},$$

$$= \cdot 4342 \quad ,, \quad m = \frac{\mu}{3},$$

$$= \cdot 3287 \quad ,, \quad m = \frac{\mu}{4}.$$

The tangential force may be calculated in the same way, it is

$$T = \frac{1}{\mu} R \{MA \sin (ms + a) + NB \sin (ms + \beta)\} \dots\dots\dots\dots (10).$$

The normal force is

$$Z = -\frac{1}{\mu} RJC \cos (ms + \gamma) \dots\dots\dots\dots\dots (11).$$

6. We have found the expressions for the forces which act upon each member of a system of equal satellites which originally formed a uniform ring, but are now affected with displacements depending on circular functions. If these displacements can be propagated round the ring in the form of waves with the velocity $\frac{m}{n}$, the quantities a, β, and γ will depend on t, and the complete expressions will be

$$\left. \begin{array}{l} \rho = A \cos (ms + nt + a) \\ \sigma = B \sin (ms + nt + \beta) \\ \zeta = C \cos (ms + n't + \gamma) \end{array} \right\} \dots\dots\dots\dots\dots (12).$$

Let us find in what cases expressions such as these will be true, and what will be the result when they are not true.

Let the position of a satellite at any time be determined by the values of r, ϕ, and ζ, where r is the radius vector reduced to the plane of reference, ϕ the angle of position measured on that plane, and ζ the distance from it. The equations of motion will be

$$\left. \begin{array}{l} r \left(\dfrac{d\phi}{dt}\right)^2 - \dfrac{d^2r}{dt^2} = S \dfrac{1}{r^2} + P \\[2mm] 2 \dfrac{dr}{dt} \dfrac{d\phi}{dt} + r \dfrac{d^2\phi}{dt^2} = T \\[2mm] \dfrac{d^2\zeta}{dt^2} = -S \dfrac{\zeta}{r^3} + Z \end{array} \right\} \dots\dots\dots\dots (13).$$

If we substitute the value of ζ in the third equation and remember that r is nearly $= 1$, we find

$$n'^2 = S + \frac{1}{\mu} RJ \dots\dots\dots\dots\dots (14).$$

As this expression is necessarily positive, the value of n' is always real, and the disturbances normal to the plane of the ring can always be propagated as waves, and therefore can never be the cause of instability. We therefore confine our attention to the motion in the plane of the ring as deduced from the two former equations.

Putting $r = 1 + \rho$ and $\phi = \omega t + s + \sigma$, and omitting powers and products of ρ, σ and their differential coefficients,

$$\left. \begin{aligned} \omega^2 + \omega^2\rho + 2\omega\,\frac{d\sigma}{dt} - \frac{d^2\rho}{dt^2} = S - 2S\rho + P \\ 2\omega\,\frac{d\rho}{dt} + \frac{d^2\sigma}{dt^2} = T \end{aligned} \right\} \quad \ldots\ldots\ldots\ldots\ldots (15).$$

Substituting the values of ρ and σ as given above, these equations become

$$\omega^2 - S - \frac{1}{\mu} RK + \left(\omega^2 + 2S - \frac{1}{\mu} RL + n^2\right) A \, \cos\,(ms + nt + a)$$

$$+ \left(2\omega n + \frac{1}{\mu} RM\right) B \, \cos\,(ms + nt + \beta) = 0 \ldots\ldots\ldots (16),$$

$$\left(2\omega n + \frac{1}{\mu} RM\right) A \, \sin\,(ms + nt + a) + \left(n^2 + \frac{1}{\mu} RN\right) B \, \sin\,(ms + nt + \beta) = 0 \ldots(17).$$

Putting for $(ms + nt)$ any two different values, we find from the second equation (17)

$$a = \beta \ldots\ldots\ldots\ldots\ldots\ldots\ldots\ldots\ldots\ldots\ldots\ldots\ldots(18),$$

and

$$\left(2\omega n + \frac{1}{\mu} RM\right) A + \left(n^2 + \frac{1}{\mu} RN\right) B = 0 \ldots\ldots\ldots\ldots\ldots (19),$$

and from the first (16) $\left(\omega^2 + 2S - \frac{1}{\mu} RL + n^2\right) A + \left(2\omega n + \frac{1}{\mu} RM\right) B = 0 \ldots\ldots\ldots (20)$,

and

$$\omega^2 - S - \frac{1}{\mu} RK = 0 \ldots\ldots\ldots\ldots\ldots\ldots\ldots\ldots\ldots(21).$$

Eliminating A and B from these equations, we get

$$n^4 - \left\{3\omega^2 - 2S + \frac{1}{\mu} R\,(L - N)\right\} n^2$$

$$- 4\omega\,\frac{1}{\mu} RMn + \left(\omega^2 + 2S - \frac{1}{\mu} RL\right)\frac{1}{\mu} RN - \frac{1}{\mu^2} R^2M^2 = 0 \ldots\ldots\ldots (22),$$

a biquadratic equation to determine n.

For every *real* value of n there are terms in the expressions for ρ and σ of the form

$$A \, \cos\,(ms + nt + a).$$

For every *pure impossible* root of the form $\pm\sqrt{-1}n'$ there are terms of the forms

$$A\epsilon^{\pm n't}\cos\left(ms+a\right).$$

Although the negative exponential coefficient indicates a continually diminishing displacement which is consistent with stability, the positive value which necessarily accompanies it indicates a continually increasing disturbance, which would completely derange the system in course of time.

For every mixed root of the form $\pm\sqrt{-1}n'+n$, there are terms of the form

$$A\epsilon^{\pm n't}\cos\left(ms+nt+a\right).$$

If we take the positive exponential, we have a series of m waves travelling with velocity $\dfrac{n}{m}$ and increasing in amplitude with the coefficient $\epsilon^{+n't}$. The negative exponential gives us a series of m waves gradually dying away, but the negative exponential cannot exist without the possibility of the positive one having a finite coefficient, so that it is necessary for the stability of the motion that the four values of n be all real, and none of them either impossible quantities or the sums of possible and impossible quantities.

We have therefore to determine the relations among the quantities K, L, M, N, R, S, that the equation

$$n^4-\left\{S+\frac{1}{\mu}\,R\left(3K+L-N\right)\right\}n^2$$

$$-4\omega\frac{1}{\mu}RMn+\left\{3S+\frac{1}{\mu}R\left(K-L\right)\right\}\frac{1}{\mu}RN-\frac{1}{\mu^2}R^2M^2=U=0$$

may have four real roots.

7. In the first place, U is positive, when n is a large enough quantity, whether positive or negative.

It is also positive when $n=0$, provided S be large, as it must be, compared with $\dfrac{1}{\mu}RL$, $\dfrac{1}{\mu}RM$ and $\dfrac{1}{\mu}RN$.

If we can now find a positive and a negative value of n for which U is negative, there must be four real values of n for which $U=0$, and the four roots will be real.

Now if we put $n = \pm \sqrt{\tfrac{1}{2}} \sqrt{S}$,

$$U = -\tfrac{1}{4}S^2 + \tfrac{1}{2}\,\frac{1}{\mu}\,R\,(7N \pm 4\sqrt{2}M - L - 3K)\,S + \frac{1}{\mu^2}\,R^2\,(KN - LN - M^2),$$

which is negative if S be large compared to R.

So that a ring of satellites can always be rendered stable by increasing the mass of the central body and the angular velocity of the ring.

The values of L, M, and N depend on m, the number of undulations in the ring. When $m = \dfrac{\mu}{2}$, the values of L and N will be at their maximum and $M = 0$. If we determine the relation between S and R in this case so that the system may be stable, the stability of the system for every other displacement will be secured.

8. To find the mass which must be given to the central body in order that a ring of satellites may permanently revolve round it.

We have seen that when the attraction of the central body is sufficiently great compared with the forces arising from the mutual action of the satellites, a permanent ring is possible. Now the forces between the satellites depend on the manner in which the displacement of each satellite takes place. The conception of a perfectly arbitrary displacement of all the satellites may be rendered manageable by separating it into a number of partial displacements depending on periodic functions. The motions arising from these small displacements will take place independently, so that we have to consider only one at a time.

Of all these displacements, that which produces the greatest disturbing forces is that in which consecutive satellites are oppositely displaced, that is, when $m = \dfrac{\mu}{2}$, for then the nearest satellites are displaced so as to increase as much as possible the effects of the displacement of the satellite between them. If we make μ a large quantity, we shall have

$$\Sigma \frac{\sin^2 m\theta \cos^2 \theta}{\sin^3 \theta} = \frac{\mu^3}{\pi^3}\,(1 + 3^{-3} + 5^{-3} + \&\text{c.}) = \frac{\mu^3}{\pi^3}\,(1\cdot 0518),$$

$$L = \frac{\mu^3}{\pi^3}\,\cdot 5259, \qquad M = 0, \qquad N = 2L, \qquad K \text{ very small.}$$

Let $\dfrac{1}{\mu} RL = x$, then the equation of motion will be

$$n^4 - (S-x)\, n^2 + 2x\, (3S-x) = U = 0 \ \ldots\ldots\ldots\ldots\ldots\ldots (23).$$

The conditions of this equation having real roots are

$$S > x \ \ldots\ldots\ldots\ldots\ldots\ldots\ldots\ldots\ldots\ldots\ldots\ldots (24),$$

$$(S-x)^2 > 8x\, (3S-x) \ \ldots\ldots\ldots\ldots\ldots\ldots\ldots (25).$$

The last condition gives the equation

$$S^2 - 26Sx + 9x^2 > 0,$$

whence $\qquad\qquad S > 26{\cdot}642x, \ \ \text{or} \ \ S < 0{\cdot}351x \ \ldots\ldots\ldots\ldots\ldots\ldots (26).$

The last solution is inadmissible because S must be greater than x, so that the true condition is $\qquad S > 25{\cdot}649x,$

$$> 25{\cdot}649\, \frac{1}{\mu}\, R\, \frac{\mu^3}{\pi^3}\, {\cdot}5259,$$

$$S > {\cdot}4352 \mu^2 R \ \ldots\ldots\ldots\ldots\ldots\ldots\ldots\ldots\ldots (27).$$

So that if there were 100 satellites in the ring, then

$$S > 4352\ R$$

is the condition which must be fulfilled in order that the motion arising from every conceivable displacement may be periodic.

If this condition be not fulfilled, and if S be not sufficient to render the motion perfectly stable, then although the motion depending upon long undulations may remain stable, the short undulations will increase in amplitude till some of the neighbouring satellites are brought into collision.

9. To determine the nature of the motion when the system of satellites is of small mass compared with the central body.

The equation for the determination of n is

$$U = n^4 - \left\{\omega^2 + \frac{1}{\mu}\, R\, (2K + L - N)\right\} n^2 - 4\omega\, \frac{1}{\mu}\, RMn$$

$$+ \left\{3\omega^2 - \frac{1}{\mu}\, R\, (2K+L)\right\}\frac{1}{\mu}\, RN - \frac{1}{\mu^2}\, R^2 M^2 = 0 \ \ldots\ldots\ldots (28).$$

When R is very small we may approximate to the values of n by assuming that two of them are nearly $\pm\,\omega$, and that the other two are small.

If we put $n = \pm \omega$,

$$U = -\frac{1}{\mu} R \left(2K + L \pm 4M - 4N\right)\omega^2 + \&\mathrm{c.},$$

$$\frac{dU}{dn} = \pm 2\omega^3 + \&\mathrm{c.}$$

Therefore the corrected values of n are

$$n = \pm \left\{\omega + \frac{1}{2\mu\omega} R \left(2K + L - 4N\right)\right\} + \frac{2}{\mu\omega} RM \dots\dots\dots\dots (29).$$

The small values of n are nearly $\pm \sqrt{3\frac{1}{\mu} RN}$: correcting them in the same way, we find the approximate values

$$n = \pm \sqrt{3\frac{1}{\mu} RN} - 2\frac{1}{\mu\omega} RM \dots\dots\dots\dots\dots (30).$$

The four values of n are therefore

$$\left. \begin{aligned} n_1 &= -\omega - \frac{1}{2\mu\omega} R(2K + L - 4M - 4N) \\[2mm] n_2 &= -\sqrt{3\frac{1}{\mu} RN} - \frac{2}{\mu\omega} RM \\[2mm] n_3 &= +\sqrt{3\frac{1}{\mu} RN} - \frac{2}{\mu\omega} RM \\[2mm] n_4 &= +\omega + \frac{1}{2\mu\omega} R(2K + L + 4M - 4N) \end{aligned} \right\} \dots\dots\dots\dots (31),$$

and the complete expression for ρ, so far as it depends on terms containing ms, is therefore

$$\rho = A_1 \cos\left(ms + n_1 t + a_1\right) + A_2 \cos\left(ms + n_2 t + a_2\right)$$
$$+ A_3 \cos\left(ms + n_3 t + a_3\right) + A_4 \cos\left(ms + n_4 t + a_4\right)\dots\dots\dots\dots(32),$$

and there will be other systems, of four terms each, for every value of m in the expansion of the original disturbance.

We are now able to determine the value of σ from equations (12), (20), by putting $\beta = a$, and

$$B = -\frac{2\omega n + \frac{1}{\mu} RM}{n^2 + \frac{1}{\mu} RN} A \dots\dots\dots\dots\dots (33).$$

So that for every term of ρ of the form

$$\rho = A \cos (ms + nt + a) \quad\dots\dots\dots\dots\dots\dots\dots\dots (34),$$

there is a corresponding term in σ,

$$\sigma = - \frac{2\omega n + \frac{1}{\mu} RM}{n^2 + \frac{1}{\mu} RN} A \sin (ms + nt + a) \quad\dots\dots\dots\dots\dots (35).$$

10. Let us now fix our attention on the motion of a single satellite, and determine its motion by tracing the changes of ρ and σ while t varies and s is constant, and equal to the value of s corresponding to the satellite in question.

We must recollect that ρ and σ are measured outwards and forwards from an imaginary point revolving at distance 1 and velocity ω, so that the motions we consider are not the absolute motions of the satellite, but its motions relative to a point fixed in a revolving plane. This being understood, we may describe the motion as elliptic, the major axis being in the tangential direction, and the ratio of the axes being nearly $2\frac{\omega}{n}$, which is nearly 2 for n_1 and n_4 and is very large for n_2 and n_3.

The time of revolution is $\frac{2\pi}{n}$, or if we take a revolution of the ring as the unit of time, the time of a revolution of the satellite about its mean position is $\frac{\omega}{n}$.

The *direction* of revolution of the satellite about its mean position is in every case opposite to that of the motion of the ring.

11. The absolute motion of a satellite may be found from its motion relative to the ring by writing

$$r = 1 + \rho = 1 + A \cos (ms + nt + a),$$

$$\theta = \omega t + s + \sigma = \omega t + s - 2\frac{\omega}{n} A \sin (ms + nt + a).$$

When n is nearly equal to $\pm\omega$, the motion of each satellite in space is nearly elliptic. The eccentricity is A, the longitude at epoch s, and the longitude when at the greatest distance from Saturn is for the negative value n_1

$$-\frac{1}{\mu\omega} R \left(2K+L-4M-4N\right) t + \left(m+1\right) s + a,$$

and for the positive value n_4

$$-\frac{1}{\mu\omega} R \left(2K+L+4M-4N\right) t - \left(m+1\right) s - a.$$

We must recollect that in all cases the quantity within brackets is negative, so that the major axis of the ellipse travels forwards in both cases. The chief difference between the two cases lies in the arrangement of the major axes of the ellipses of the different satellites. In the first case as we pass from one satellite to the next in front the axes of the two ellipses lie in the same order. In the second case the particle in front has its major axis behind that of the other. In the cases in which n is small the radius vector of each satellite increases and diminishes during a periodic time of several revolutions. This gives rise to an inequality, in which the tangential displacement far exceeds the radial, as in the case of the *annual equation* of the Moon.

12. Let us next examine the condition of the ring of satellites at a given instant. We must therefore fix on a particular value of t and trace the changes of ρ and σ for different values of s.

From the expression for ρ we learn that the satellites form a wavy line, which is furthest from the centre when $(ms + nt + a)$ is a multiple of 2π, and nearest to the centre for intermediate values.

From the expression for σ we learn that the satellites are sometimes in advance and sometimes in the rear of their mean position, so that there are places where the satellites are crowded together, and others where they are drawn asunder. When n is positive, B is of the opposite sign to A, and the crowding of the satellites takes place when they are furthest from the centre. When n is negative, the satellites are separated most when furthest from the centre, and crowded together when they approach it.

The form of the ring at any instant is therefore that of a string of beads forming a re-entering curve, nearly circular, but with a small variation of distance from the centre recurring m times, and forming m regular waves of transverse displacement at equal intervals round the circle. Besides these, there are waves of condensation and rarefaction, the effect of longitudinal displacement. When n is positive the points of greatest distance from the centre are points of greatest condensation, and when n is negative they are points of greatest rarefaction.

13. We have next to determine the velocity with which these waves of disturbance are propagated round the ring. We fixed our attention on a particular satellite by making s constant, and on a particular instant by making t constant, and thus we determined the motion of a satellite and the form of the ring. We must now fix our attention on a *phase* of the motion, and this we do by making ρ or σ constant. This implies

$$ms + nt + a = \text{constant},$$

$$\frac{ds}{dt} = -\frac{n}{m}.$$

So that the particular phase of the disturbance travels round the ring with an angular velocity $= -\dfrac{n}{m}$ relative to the ring itself. Now the ring is revolving in space with the velocity ω, so that the angular velocity of the wave in space is

$$\varpi = \omega - \frac{n}{m} \quad\dotfill(36).$$

Thus each satellite moves in an ellipse, while the general aspect of the ring is that of a curve of m waves revolving with velocity ϖ. This, however, is only the part of the whole motion, which depends on a single term of the solution. In order to understand the general solution we must shew how to determine the whole motion from the state of the ring at a given instant.

14. *Given the position and motion of every satellite at any one time, to calculate the position and motion of every satellite at any other time, provided that the condition of stability is fulfilled.*

The position of any satellite may be denoted by the values of ρ and σ for that satellite, and its velocity and direction of motion are then indicated by the values of $\dfrac{d\rho}{dt}$ and $\dfrac{d\sigma}{dt}$ at the given instant.

These four quantities may have for each satellite any four arbitrary values, as the position and motion of each satellite are independent of the rest, at the beginning of the motion.

Each of these quantities is therefore a perfectly arbitrary function of s, the mean angular position of the satellite in the ring.

But any function of s from $s = 0$ to $s = 2\pi$, however arbitrary or discontinuous, can be expanded in a series of terms of the form $A \cos (s + a) + A' \cos (2s + a') +$ &c. See § 3.

Let each of the four quantities ρ, $\dfrac{d\rho}{dt}$, σ, $\dfrac{d\sigma}{dt}$ be expressed in terms of such a series, and let the terms in each involving ms be

$$\rho = E \cos (ms + e) \dots\dots\dots\dots\dots\dots (37),$$

$$\frac{d\rho}{dt} = F \cos (ms + f) \dots\dots\dots\dots\dots\dots (38),$$

$$\sigma = G \cos (ms + g) \dots\dots\dots\dots\dots\dots (39),$$

$$\frac{d\sigma}{dt} = H \cos (ms + h) \dots\dots\dots\dots\dots\dots (40).$$

These are the parts of the values of each of the four quantities which are capable of being expressed in the form of periodic functions of ms. It is evident that the eight quantities E, F, G, H, e, f, g, h, are all independent and arbitrary.

The next operation is to find the values of L, M, N, belonging to disturbances in the ring whose index is m [see equation (8)], to introduce these values into equation (28), and to determine the four values of n, (n_1, n_2, n_3, n_4).

This being done, the expression for ρ is that given in equation (32), which contains eight arbitrary quantities $(A_1, A_2, A_3, A_4, a_1, a_2, a_3, a_4)$.

Giving t its original value in this expression, and equating it to $E \cos (ms + e)$, we get an equation which is equivalent to two. For, putting $ms = 0$, we have

$$A_1 \cos a_1 + A_2 \cos a_2 + A_3 \cos a_3 + A_4 \cos a_4 = E \cos e \dots\dots\dots (41).$$

And putting $ms = \dfrac{\pi}{2}$, we have another equation

$$A_1 \sin a_1 + A_2 \sin a_2 + A_3 \sin a_3 + A_4 \sin a_4 = E \sin e \dots\dots\dots\dots (42).$$

Differentiating (32) with respect to t, we get two other equations

$$- A_1 n_1 \sin a - \&\text{c.} = F \cos f \dots\dots\dots\dots\dots\dots\dots (43),$$

$$A_1 n_1 \cos a + \&\text{c.} = F \sin f \dots\dots\dots\dots\dots\dots\dots (44).$$

Bearing in mind that B_1, B_2, &c. are connected with A_1, A_2, &c. by equation (33), and that B is therefore proportional to A, we may write $B = A\beta$, where

$$\beta = - \frac{2\omega n + \dfrac{1}{\mu} RM}{n^2 + \dfrac{1}{\mu} RN},$$

β being thus a function of n and a known quantity.

The value of σ then becomes at the epoch

$$\sigma = A_1 \beta_1 \sin (ms + a_1) + \&\text{c.} = G \cos (ms + g),$$

from which we obtain the two equations

$$A_1 \beta_1 \sin a_1 + \&\text{c.} = G \cos g \dots\dots\dots\dots\dots\dots\dots(45),$$

$$A_1 \beta_1 \cos a_1 + \&\text{c.} = - G \sin g \dots\dots\dots\dots\dots\dots\dots(46).$$

Differentiating with respect to t, we get the remaining equations

$$A_1 \beta_1 n_1 \cos a_1 + \&\text{c.} = H \cos h \dots\dots\dots\dots\dots\dots (47),$$

$$A_1 \beta_1 n_1 \sin a_1 + \&\text{c.} = H \sin h \dots\dots\dots\dots\dots\dots (48).$$

We have thus found eight equations to determine the eight quantities A_1, &c. and a_1, &c. To solve them, we may take the four in which $A_1 \cos a_1$, &c. occur, and treat them as simple equations, so as to find $A_1 \cos a_1$, &c. Then taking those in which $A_1 \sin a_1$, &c. occur, and determining the values of those quantities, we can easily deduce the value of A_1 and a_1, &c. from these.

We now know the amplitude and phase of each of the four waves whose index is m. All other systems of waves belonging to any other index must be treated in the same way, and since the original disturbance, however arbitrary, can be broken up into periodic functions of the form of equations (37—40), our solution is perfectly general, and applicable to every possible disturbance of a ring fulfilling the condition of stability (27).

15. We come next to consider the effect of an external disturbing force, due either to the irregularities of the planet, the attraction of satellites, or the motion of waves in other rings.

All disturbing forces of this kind may be expressed in series of which the general term is

$$A \cos (vt + ms + a),$$

where v is an angular velocity and m a whole number.

Let $P \cos (ms + vt + p)$ be the central part of the force, acting inwards, and $Q \sin (ms + vt + q)$ the tangential part, acting forwards. Let $\rho = A \cos (ms + vt + a)$ and $\sigma = B \sin (ms + vt + \beta)$, be the terms of ρ and σ which depend on the external disturbing force. These will simply be added to the terms depending on the original disturbance which we have already investigated, so that the complete expressions for ρ and σ will be as general as before. In consequence of the additional forces and displacements, we must add to equations (16) and (17), respectively, the following terms:

$$\{3\omega^2 - \frac{1}{\mu} R (2K + L) + v^2\} A \cos (ms + vt + a)$$

$$+ \left(2\omega v + \frac{1}{\mu} RM\right) B \cos (ms + vt + \beta) - P \cos (ms + vt + p) = 0 \ldots \ldots (49).$$

$$\left(2\omega v + \frac{1}{\mu} RM\right) A \sin (ms + vt + a)$$

$$+ \left(v^2 + \frac{1}{\mu} RN\right) B \sin (ms + vt + \beta) + Q \sin (ms + vt + q) = 0 \ldots \ldots \ldots (50).$$

Making $ms + vt = 0$ in the first equation and $\frac{\pi}{2}$ in the second,

$$\{3\omega^2 - \frac{1}{\mu} R (2K + L) + v^2\} A \cos a + \left(2\omega v + \frac{1}{\mu} RM\right) B \cos \beta - P \cos p = 0 \ldots \ldots (51).$$

$$\left(2\omega v + \frac{1}{\mu} RM\right) A \cos a + \left(v^2 + \frac{1}{\mu} RN\right) B \cos \beta + Q \cos q = 0 \ldots \ldots (52).$$

Then if we put

$$U' = v^4 - \{\omega^2 + \frac{1}{\mu} R (2K + L - N)\} v^2 - 4 \frac{\omega}{\mu} RMv$$

$$+ \{3\omega^2 - \frac{1}{\mu} R (2K + L)\} \frac{1}{\mu} RN - \frac{1}{\mu^2} R^2 M^2 \ldots \ldots \ldots (53),$$

we shall find the value of $A \cos a$ and $B \cos \beta$;

$$A \cos a = \frac{v^2 + \frac{1}{\mu} RN}{U'} P \cos p + \frac{2\omega v + \frac{1}{\mu} RM}{U'} Q \cos q \;\ldots\ldots\ldots\ldots (54).$$

$$B \cos \beta = -\frac{2\omega v + \frac{1}{\mu} RM}{U'} P \cos p - \frac{v^2 + 3\omega^2 - \frac{1}{\mu} R(K+L)}{U'} Q \cos q \ldots\ldots(55).$$

Substituting sines for cosines in equations (51), (52), we may find the values of $A \sin a$ and $B \sin \beta$.

Now U' is precisely the same function of v that U is of n, so that if v coincides with one of the four values of n, U' will vanish, the coefficients A and B will become infinite, and the ring will be destroyed. The disturbing force is supposed to arise from a revolving body, or an undulation of any kind which has an angular velocity $-\dfrac{v}{m}$ relatively to the ring, and therefore an absolute angular velocity $= \omega - \dfrac{v}{m}$.

If then the absolute angular velocity of the disturbing body is exactly or nearly equal to the absolute angular velocity of any of the free waves of the ring, that wave will increase till the ring be destroyed.

The velocities of the free waves are nearly

$$\omega \left(1 + \frac{1}{m}\right), \; \omega + \frac{1}{m} \sqrt{3 \frac{1}{\mu} RN}, \; \omega - \frac{1}{m} \sqrt{3 \frac{1}{\mu} RN}, \; \text{and } \omega \left(1 - \frac{1}{m}\right) \ldots\ldots (56).$$

When the angular velocity of the disturbing body is greater than that of the first wave, between those of the second and third, or less than that of the fourth, U' is positive. When it is between the first and second, or between the third and fourth, U' is negative.

Let us now simplify our conception of the disturbance by attending to the central force only, and let us put $p = 0$, so that P is a maximum when $ms + vt$ is a multiple of 2π. We find in this case $a = 0$, and $\beta = 0$. Also

$$A = \frac{v^2 + \frac{1}{\mu} RN}{U'} P \;\ldots\ldots\ldots\ldots\ldots\ldots\ldots\ldots\ldots (57),$$

$$B = -\frac{2\omega v + \frac{1}{\mu} RM}{U'} P \;\ldots\ldots\ldots\ldots\ldots\ldots\ldots (58).$$

When U' is positive, A will be of the same sign as P, that is, the parts of the ring will be furthest from the centre where the disturbing force towards the centre is greatest. When U' is negative, the contrary will be the case.

When v is positive, B will be of the opposite sign to A, and the parts of the ring furthest from the centre will be most crowded. When v is negative, the contrary will be the case.

Let us now attend only to the tangential force, and let us put $q = 0$. We find in this case also $a = 0$, $\beta = 0$,

$$A = \frac{2\omega v + \frac{1}{\mu} RM}{U'} Q \dots\dots\dots\dots\dots\dots\dots (59),$$

$$B = - \frac{v^2 + 3\omega^2 - \frac{1}{\mu} R(K+L)}{U'} Q \dots\dots\dots\dots (60).$$

The tangential displacement is here in the same or in the opposite direction to the tangential force, according as U' is negative or positive. The crowding of satellites is at the points farthest from or nearest to Saturn according as v is positive or negative.

16.   The effect of any disturbing force is to be determined in the following manner. The disturbing force, whether radial or tangential, acting on the ring may be conceived to vary from one satellite to another, and to be different at different times. It is therefore a perfectly arbitrary function of s and t.

Let Fourier's method be applied to the general disturbing force so as to divide it up into terms depending on periodic functions of s, so that each term is of the form $F(t) \cos(ms + a)$, where the function of t is still perfectly arbitrary.

But it appears from the general theory of the permanent motions of the heavenly bodies that they may all be expressed by periodic functions of t arranged in series. Let vt be the argument of one of these terms, then the corresponding term of the disturbance will be of the form

$$P \cos(ms + vt + a).$$

This term of the disturbing force indicates an alternately positive and negative action, disposed in m waves round the ring, completing its period relatively to each particle in the time $\frac{2\pi}{v}$, and travelling as a wave among

the particles with an angular velocity $-\frac{v}{m}$, the angular velocity relative to fixed

space being of course $\omega - \frac{v}{m}$. The whole disturbing force may be split up into

terms of this kind.

17. Each of these elementary disturbances will produce its own wave in the ring, independent of those which belong to the ring itself. This new wave, due to external disturbance, and following different laws from the natural waves of the ring, is called the *forced wave*. The angular velocity of the forced wave is the same as that of the disturbing force, and its maxima and minima coincide with those of the force, but the extent of the disturbance and its direction depend on the comparative velocities of the forced wave and the four natural waves.

When the velocity of the forced wave lies between the velocities of the two middle free waves, or is greater than that of the swiftest, or less than that of the slowest, then the radial displacement due to a radial disturbing force is in the same direction as the force, but the tangential displacement due to a tangential disturbing force is in the opposite direction to the force.

The radial force therefore in this case produces a *positive forced wave,* and the tangential force a *negative forced wave.*

When the velocity of the forced wave is either between the velocities of the first and second free waves, or between those of the third and fourth, then the radial disturbance produces a forced wave in the contrary direction to that in which it acts, or a negative wave, and the tangential force produces a positive wave.

The coefficient of the forced wave changes sign whenever its velocity passes through the value of any of the velocities of the free waves, but it does so by becoming infinite, and not by vanishing, so that when the angular velocity very nearly coincides with that of a free wave, the forced wave becomes very great, and if the velocity of the disturbing force were made exactly equal to that of a free wave, the coefficient of the forced wave would become infinite. In such a case we should have to readjust our approximations, and to find whether such a coincidence might involve a physical impossibility.

The forced wave which we have just investigated is that which would maintain itself in the ring, supposing that it had been set agoing at the commencement of the motion. It is in fact the form of dynamical equilibrium of the ring under the influence of the given forces. In order to find the actual motion of the ring we must combine this forced wave with all the free waves, which go on independently of it, and in this way the solution of the problem becomes perfectly complete, and we can determine the whole motion under any given initial circumstances, as we did in the case where no disturbing force acted.

For instance, if the ring were perfectly uniform and circular at the instant when the disturbing force began to act, we should have to combine with the constant forced wave a system of four free waves so disposed, that at the given epoch, the displacements due to them should exactly neutralize those due to the forced wave. By the combined effect of these four free waves and the forced one the whole motion of the ring would be accounted for, beginning from its undisturbed state.

The disturbances which are of most importance in the theory of Saturn's rings are those which are produced in one ring by the action of attractive forces arising from waves belonging to another ring.

The effect of this kind of action is to produce in each ring, besides its own four free waves, four forced waves corresponding to the free waves of the other ring. There will thus be eight waves in each ring, and the corresponding waves in the two rings will act and react on each other, so that, strictly speaking, every one of the waves will be in some measure a forced wave, although the system of eight waves will be the free motion of the two rings taken together. The theory of the mutual disturbance and combined motion of two concentric rings of satellites requires special consideration.

18. On the motion of a ring of satellites when the conditions of stability are not fulfilled.

We have hitherto been occupied with the case of a ring of satellites, the stability of which was ensured by the smallness of mass of the satellites compared with that of the central body. We have seen that the statically unstable condition of each satellite between its two immediate neighbours may be compensated by the dynamical effect of its revolution round the planet, and a planet of sufficient mass can not only direct the motion of such satellites round its own body, but can likewise exercise an influence over their relations to each other, so as to overrule their natural tendency to crowd together, and distribute and preserve them in the form of a ring.

We have traced the motion of each satellite, the general shape of the disturbed ring, and the motion of the various waves of disturbance round the ring, and determined the laws both of the natural or free waves of the ring, and of the forced waves, due to extraneous disturbing forces.

We have now to consider the cases in which such a permanent motion of the ring is impossible, and to determine the mode in which a ring, originally regular, will break up, in the different cases of instability.

The equation from which we deduce the conditions of stability is—

$$U = n^4 - \left\{\omega^2 + \frac{1}{\mu} R\left(2K + L - N\right)\right\} n^2 - 4\omega \frac{1}{\mu} RMn$$
$$+ \left\{3\omega^2 - \frac{1}{\mu} R\left(2K + L\right)\right\} \frac{1}{\mu} RN - \frac{1}{\mu^2} R^2 M^2 = 0.$$

The quantity, which, in the critical cases, determines the nature of the roots of this equation, is N. The quantity M in the third term is always small compared with L and N when m is large, that is, in the case of the dangerous short waves. We may therefore begin our study of the critical cases by leaving out the third term. The equation then becomes a quadratic in n^2, and in order that all the values of n may be real, both values of n^2 must be real and positive.

The condition of the values of n^2 being real is

$$\omega^4 + \omega^2 \frac{1}{\mu} R\left(4K + 2L - 14N\right) + \frac{1}{\mu^2} R^2 \left(2K + L + N\right)^2 > 0 \ldots\ldots(61),$$

which shews that ω^2 must either be about 14 times at least smaller, or about 14 times at least greater, than quantities like $\frac{1}{\mu} RN$.

That both values of n^2 may be positive, we must have

$$\left.\begin{array}{l}\omega^2 + \frac{1}{\mu} R\left(2K + L - N\right) > 0 \\ \left\{3\omega^2 - \frac{1}{\mu} R\left(2K + L\right)\right\} \frac{1}{\mu} RN > 0\end{array}\right\} \ldots\ldots\ldots\ldots (62).$$

We must therefore take the larger value of ω^2, and also add the condition that N be positive.

We may therefore state roughly, that, to ensure stability, $\dfrac{RN}{\mu}$, the coefficient of tangential attraction, must lie between zero and $\tfrac{1}{14}\omega^2$. If the quantity be negative, the two *small* values of n will become *pure impossible* quantities. If it exceed $\tfrac{1}{14}\omega^2$, *all* the values of n will take the form of mixed impossible quantities.

If we write x for $\dfrac{1}{\mu} RN$, and omit the other disturbing forces, the equation becomes
$$U = n^4 - (\omega^2 - x)\, n^2 + 3\omega^2 x = 0 \quad\ldots\ldots\ldots\ldots\ldots\ldots \text{(63)},$$
whence
$$n^2 = \tfrac{1}{2}(\omega^2 - x) \pm \tfrac{1}{2}\sqrt{\omega^4 - 14\omega^2 x + x^2} \quad\ldots\ldots\ldots\ldots \text{(64)}.$$

If x be small, two of the values of n are nearly $\pm\omega$, and the others are small quantities, real when x is positive and impossible when x is negative.

If x be greater than $(7 - \sqrt{48})\,\omega^2$, or $\dfrac{\omega^2}{14}$ nearly, the term under the radical becomes negative, and the value of n becomes
$$n = \pm\tfrac{1}{2}\sqrt{\sqrt{12\omega^2 x} + \omega^2 - x} \pm \tfrac{1}{2}\sqrt{-1}\,\sqrt{\sqrt{12\omega^2 x} - \omega^2 + x} \quad\ldots\ldots \text{(65)},$$
where one of the terms is a real quantity, and the other impossible. Every solution may be put under the form
$$n = p \pm \sqrt{-1}\,q \quad\ldots\ldots\ldots\ldots\ldots\ldots\ldots\ldots \text{(66)},$$
where $q = 0$ for the case of stability, $p = 0$ for the pure impossible roots, and p and q finite for the mixed roots.

Let us now adopt this general solution of the equation for n, and determine its mechanical significance by substituting for the impossible circular functions their equivalent real exponential functions.

Substituting the general value of n in equations (34), (35),
$$\rho = A\left[\cos\{ms + (p + \sqrt{-1}q)\,t + a\} + \cos\{ms + (p - \sqrt{-1}q)\,t + a\}\right] \ldots \text{(67)},$$
$$\left.\begin{aligned}
\sigma = &-A\,\frac{2\omega(p + \sqrt{-1}q)}{(p + \sqrt{-1}q)^2 + x}\,\sin\{ms + (p + \sqrt{-1}q)\,t + a\} \\
&-A\,\frac{2\omega(p - \sqrt{-1}q)}{(p - \sqrt{-1}q)^2 + x}\,\sin\{ms + (p - \sqrt{-1}q)\,t + a\}
\end{aligned}\right\} \ldots\ldots \text{(68)}.$$

Introducing the exponential notation, these values become

$$\rho = A \left(\epsilon^{qt} + \epsilon^{-qt} \right) \cos \left(ms + pt + a \right) \quad\text{............................ (69)},$$

$$\sigma = -\frac{2\omega A}{\left(p^2 + q^2\right)^2 + 2\left(p^2 - q^2\right)x + x^2} \left\{ \begin{array}{l} p\left(p^2 + q^2 + x\right)\left(\epsilon^{qt} + \epsilon^{-qt}\right) \sin\left(ms + pt + a\right) \\ + q\left(p^2 + q^2 - x\right)\left(\epsilon^{qt} - \epsilon^{-qt}\right) \cos\left(ms + pt + a\right) \end{array} \right\} \ \dots (70).$$

We have now obtained a solution free from impossible quantities, and applicable to every case.

When $q = 0$, the case becomes that of real roots, which we have already discussed. When $p = 0$, we have the case of pure impossible roots arising from the negative values of n^2. The solutions corresponding to these roots are

$$\rho = A \left(\epsilon^{qt} + \epsilon^{-qt} \right) \cos \left(ms + a \right) \quad\text{.......................... (71)}.$$

$$\sigma = -\frac{2\omega q A}{q^2 - x} \left(\epsilon^{qt} - \epsilon^{-qt} \right) \cos \left(ms + a \right) \quad\text{.................. (72)}.$$

The part of the coefficient depending on ϵ^{-qt} diminishes indefinitely as the time increases, and produces no marked effect. The other part, depending on ϵ^{qt}, increases in a geometrical proportion as the time increases arithmetically, and so breaks up the ring. In the case of x being a small negative quantity, q^2 is nearly $3x$, so that the coefficient of σ becomes

$$-3\frac{\omega}{q}A.$$

It appears therefore that the motion of each particle is either outwards and backwards or inwards and forwards, but that the tangential part of the motion greatly exceeds the normal part.

It may seem paradoxical that a tangential force, acting *towards* a position of equilibrium, should produce instability, while a small tangential force *from* that position ensures stability, but it is easy to trace the destructive tendency of this apparently conservative force.

Suppose a particle slightly in front of a crowded part of the ring, then if x is negative there will be a tangential force pushing it forwards, and this force will cause its distance from the planet to increase, its angular velocity to diminish, and the particle itself to fall back on the crowded part, thereby increasing the irregularity of the ring, till the whole ring is broken up. In the same way it may be shewn that a particle *behind* a crowded part will be pushed into it. The only force which could preserve the ring from the effect of this action, is one which would prevent the particle from receding from the planet under the influence of the tangential force, or at least prevent the diminution of angular velocity. The transversal force of attraction of the ring is of this kind, and acts in the right direction, but it can never be of sufficient magnitude to have the required effect. In fact the thing to be done is to render the last term of the equation in n^2 positive when N is negative, which requires

$$\frac{1}{\mu} R\,(2K+L) > 3\omega^2,$$

and this condition is quite inconsistent with any constitution of the ring which fulfils the other condition of stability which we shall arrive at presently.

We may observe that the waves belonging to the two real values of n, $\pm\,\omega$, must be conceived to be travelling round the ring during the whole time of its breaking up, and conducting themselves like ordinary waves, till the excessive irregularities of the ring become inconsistent with their uniform propagation.

The irregularities which depend on the exponential solutions do not travel round the ring by propagation among the satellites, but remain among the same satellites which first began to move irregularly.

We have seen the fate of the ring when x is negative. When x is small we have two small and two large values of n, which indicate regular waves, as we have already shewn. As x increases, the small values of n increase, and the large values diminish, till they meet and form a pair of positive and a pair of negative equal roots, having values nearly $\pm\,^.68\omega$. When x becomes greater than about $\frac{1}{14}\omega^2$, then all the values of n become impossible, of the form $p+\sqrt{-1}q$, q being small when x first begins to exceed its limits, and p being nearly $\pm\,^.68\omega$.

The values of ρ and σ indicate periodic inequalities having the period $\dfrac{2\pi}{p}$, but increasing in amplitude at a rate depending on the exponential ϵ^{qt}. At the beginning of the motion the oscillations of the particles are in ellipses as in the case of stability, having the ratio of the axes about 1 in the normal direction to 3 in the tangential direction. As the motion continues, these ellipses increase in magnitude, and another motion depending on the second term of σ is combined with the former, so as to increase the ellipticity of the oscillations and to turn the major axis into an inclined position, so that its fore end points a little inwards, and its hinder end a little outwards. The oscillations of each particle round its mean position are therefore in ellipses, of which both axes increase continually while the eccentricity increases, and the major axis becomes slightly inclined to the tangent, and this goes on till the ring is destroyed. In the mean time the irregularities of the ring do not remain among the same set of particles as in the former case, but travel round the ring with a relative angular velocity $-\frac{p}{m}$. Of these waves there are four, two travelling forwards among the satellites, and two travelling backwards. One of each of these pairs depends on a negative value of q, and consists of a wave whose amplitude continually decreases. The other depends on a positive value of q, and is the destructive wave whose character we have just described.

19. We have taken the case of a ring composed of equal satellites, as that with which we may compare other cases in which the ring is constructed of loose materials differently arranged.

In the first place let us consider what will be the conditions of a ring composed of satellites of unequal mass. We shall find that the motion is of the same kind as when the satellites are equal.

For by arranging the satellites so that the smaller satellites are closer together than the larger ones, we may form a ring which will revolve uniformly about Saturn, the resultant force on each satellite being just sufficient to keep it in its orbit.

To determine the stability of this kind of motion, we must calculate the disturbing forces due to any given displacement of the ring. This calculation will be more complicated than in the former case, but will lead to results of the same general character. Placing these forces in the equations of motion, we shall find a solution of the same general character as in the former case, only instead of regular waves of displacement travelling round the ring, each wave will be split and reflected when it comes to irregularities in the chain of satellites. But if the condition of stability for every kind of wave be fulfilled, the motion of each satellite will consist of small oscillations about its position of dynamical equilibrium, and thus, on the whole, the ring will of itself assume the arrangement necessary for the continuance of its motion, if it be originally in a state not very different from that of equilibrium.

20. We now pass to the case of a ring of an entirely different construction. It is possible to conceive of a quantity of matter, either solid or liquid, not collected into a continuous mass, but scattered thinly over a great extent of space, and having its motion regulated by the gravitation of its parts to each other, or towards some dominant body. A shower of rain, hail, or cinders is a familiar illustration of a number of unconnected particles in motion; the visible stars, the milky way, and the resolved nebulæ, give us instances of a similar scattering of bodies on a larger scale. In the terrestrial instances we see the motion plainly, but it is governed by the attraction of the earth, and retarded by the resistance of the air, so that the mutual attraction of the parts is completely masked. In the celestial cases the distances are so enormous, and the time during which they have been observed so short, that we can perceive no motion at all. Still we are perfectly able to conceive of a collection of particles of small size compared with the distances between them, acting upon one another only by the attraction of gravitation, and revolving round a central body. The average density of such a system may be smaller than that of the rarest gas, while the particles themselves may be of great density; and the appearance from a distance will be that of a cloud of vapour, with this difference, that as the space between the particles is empty, the rays of light will pass through the system without being refracted, as they would have been if the system had been gaseous.

Such a system will have an *average density* which may be greater in some places than others. The resultant attraction will be towards places of greater average density, and thus the density of those places will be increased so as to increase the irregularities of density. The system will therefore be statically unstable, and nothing but motion of some kind can prevent the particles from forming agglomerations, and these uniting, till all are reduced to one solid mass.

We have already seen how dynamical stability can exist where there is statical instability in the case of a row of particles revolving round a central body. Let us now conceive a cloud of particles forming a ring of nearly uniform density revolving about a central body. There will be a primary effect of inequalities in density tending to draw particles towards the denser parts of the ring, and this will elicit a secondary effect, due to the motion of revolution, tending in the contrary direction, so as to restore the rings to uniformity. The relative magnitude of these two opposing forces determines the destruction or preservation of the ring.

To calculate these effects we must begin with the statical problem:—To determine the forces arising from the given displacements of the ring.

The longitudinal force arising from longitudinal displacements is that which has most effect in determining the stability of the ring. In order to estimate its limiting value we shall solve a problem of a simpler form.

21. An infinite mass, originally of uniform density k, has its particles displaced by a quantity ξ parallel to the axis of x, so that $\xi = A \cos mx$, to determine the attraction on each particle due to this displacement.

The density at any point will differ from the original density by a quantity k', so that

$$(k + k') \, (dx + d\xi) = k \, dx \dotfill (73),$$

$$k' = -k \frac{d\xi}{dx} = Akm \sin mx \dotfill (74).$$

The potential at any point will be $V + V'$, where V is the original potential, and V' depends on the displacement only, so that

$$\frac{d^2 V'}{dx^2} + \frac{d^2 V'}{dy^2} + \frac{d^2 V'}{dz^2} + 4\pi k' = 0 \dotfill (75).$$

Now V' is a function of x only, and therefore,

$$V' = 4\pi A k \, \frac{1}{m} \sin mx \dotfill (76),$$

and the longitudinal force is found by differentiating V' with respect to x.

$$X = \frac{dV'}{dx} = 4\pi k A \cos mx = 4\pi k \xi \dotfill (77).$$

Now let us suppose this mass not of infinite extent, but of finite section parallel to the plane of yz. This change amounts to cutting off all portions of the mass beyond a certain boundary. Now the effect of the portion so cut off upon the longitudinal force depends on the value of m. When m is large, so that the wave-length is small, the effect of the external portion is insensible, so that the longitudinal force due to short waves is not diminished by cutting off a great portion of the mass.

22. Applying this result to the case of a ring, and putting s for x, and σ for ξ we have

$$\sigma = A \cos ms, \text{ and } T = 4\pi k A \cos ms,$$

so that

$$\frac{1}{\mu} RN = 4\pi k,$$

when m is very large, and this is the greatest value of N.

The value of L has little effect on the condition of stability. If L and M are both neglected, that condition is

$$\omega^2 > 27{\cdot}856 \ (2\pi k) \dots\dots\dots\dots\dots\dots\dots\dots (78),$$

and if L be as much as $\frac{1}{2}N$, then

$$\omega^2 > 25{\cdot}649 \ (2\pi k) \dots\dots\dots\dots\dots\dots\dots\dots(79),$$

so that it is not important whether we calculate the value of L or not.

The condition of stability is, that the average density must not exceed a certain value. Let us ascertain the relation between the maximum density of the ring and that of the planet.

Let b be the radius of the planet, that of the ring being unity, then the mass of Saturn is $\frac{4}{3}\pi b^3 k' = \omega^2$ if k' be the density of the planet. If we assume that the radius of the ring is twice that of the planet, as Laplace has done, then $b = \frac{1}{2}$ and

$$\frac{k'}{k} = 334{\cdot}2 \text{ to } 307{\cdot}7 \dots\dots\dots\dots\dots\dots\dots\dots (80),$$

so that the density of the ring cannot exceed $\frac{1}{300}$ of that of the planet. Now Laplace has shewn that if the outer and inner parts of the ring have the same angular velocity, the ring will not hold together if the ratio of the density of the planet to that of the ring exceeds $1{\cdot}3$, so that in the first place, our ring cannot have uniform angular velocity, and in the second place, Laplace's ring cannot preserve its form, if it is composed of loose materials acting on each other only by the attraction of gravitation, and moving with the same angular velocity throughout.

23. On the forces arising from inequalities of thickness in a thin stratum of fluid of indefinite extent.

The forces which act on any portion of a continuous fluid are of two kinds, the pressures of contiguous portions of fluid, and the attractions of all portions of the fluid whether near or distant. In the case of a thin stratum of fluid, not acted on by any external forces, the pressures are due mainly to the component of the attraction which is perpendicular to the plane of the stratum. It is easy to shew that a fluid acted on by such a force will tend to assume a position of equilibrium, in which its free surface is plane ; and that any irregularities will tend to equalise themselves, so that the plane surface will be one of stable equilibrium.

It is also evident, that if we consider only that part of the attraction which is parallel to the plane of the stratum, we shall find it always directed towards the thicker parts, so that the effect of this force is to draw the fluid from thinner to thicker parts, and so to increase irregularities and destroy equilibrium.

The normal attraction therefore tends to preserve the stability of equilibrium, while the tangential attraction tends to render equilibrium unstable.

According to the nature of the irregularities one or other of these forces will prevail, so that if the extent of the irregularities is small, the normal forces will ensure stability, while, if the inequalities cover much space, the tangential forces will render equilibrium unstable, and break up the stratum into beads.

To fix our ideas, let us conceive the irregularities of the stratum split up into the form of a number of systems of waves superposed on one another, then, by what we have just said, it appears, that very short waves will disappear of themselves, and be consistent with stability, while very long waves will tend to increase in height, and will destroy the form of the stratum.

In order to determine the law according to which these opposite effects take place, we must subject the case to mathematical investigation.

Let us suppose the fluid incompressible, and of the density k, and let it be originally contained between two parallel planes, at distances $+c$ and $-c$ from that of (xy), and extending to infinity. Let us next conceive a series of imaginary planes, parallel to the plane of (yz), to be plunged into the fluid stratum at infinitesimal distances from one another, so as to divide the fluid into imaginary slices perpendicular to the plane of the stratum.

Next let these planes be displaced parallel to the axis of x according to this law—that if x be the original distance of the plane from the origin, and ξ its displacement in the direction of x,

$$\xi = A \cos mx \quad\text{......................................} (81).$$

According to this law of displacement, certain alterations will take place in the distances between consecutive planes; but since the fluid is incompressible, and of indefinite extent in the direction of y, the change of dimension must occur in the direction of z. The original thickness of the stratum was $2c$. Let its thickness at any point after displacement be $2c+2\zeta$, then we must have

$$(2c+2\zeta)\left(1+\frac{d\xi}{dx}\right)=2c \quad\ldots\ldots\ldots\ldots\ldots\ldots (82),$$

or
$$\zeta=-c\frac{d\xi}{dx}=cmA\sin mx \ldots\ldots\ldots\ldots\ldots\ldots(83).$$

Let us assume that the increase of thickness 2ζ is due to an increase of ζ at each surface; this is necessary for the equilibrium of the fluid between the imaginary planes.

We have now produced artificially, by means of these planes, a system of waves of longitudinal displacement whose length is $\frac{2\pi}{m}$ and amplitude A; and we have found that this has produced a system of waves of normal displacement on each surface, having the same length, with a height $=cmA$.

In order to determine the forces arising from these displacements, we must, in the first place, determine the potential function at any point of space, and this depends partly on the state of the fluid before displacement, and partly on the displacement itself. We have, in all cases—

$$\frac{d^2V}{dx^2}+\frac{d^2V}{dy^2}+\frac{d^2V}{dz^2}=-4\pi\rho \quad\ldots\ldots\ldots\ldots\ldots\ldots (84).$$

Within the fluid, $\rho=k$; beyond it, $\rho=0$.

Before displacement, the equation is reduced to

$$\frac{d^2V}{dz^2}=-4\pi\rho\ldots\ldots\ldots\ldots\ldots\ldots\ldots\ldots (85).$$

Instead of assuming $V=0$ at infinity, we shall assume $V=0$ at the origin, and since in this case all is symmetrical, we have

within the fluid
$$V_1=-2\pi kz^2;\quad \frac{dV_1}{dz}=-4\pi kz$$

at the bounding planes
$$V=-2\pi kc^2;\quad \frac{dV}{dz}=\mp4\pi kc \quad\left.\right\}\ldots\ldots\ldots\ldots(86);$$

beyond them
$$V_2=2\pi kc\,(\mp2z\pm c);\quad \frac{dV}{dz}=\mp4\pi kc$$

the upper sign being understood to refer to the boundary at distance $+c$, and the lower to the boundary at distance $-c$ from the origin.

Having ascertained the potential of the undisturbed stratum, we find that of the disturbance by calculating the effect of a stratum of density k and thickness ζ, spread over each surface according to the law of thickness already found. By supposing the coefficient A small enough, (as we may do in calculating the displacements on which stability depends), we may diminish the absolute thickness indefinitely, and reduce the case to that of a mere " superficial density," such as is treated of in the theory of electricity. We have here, too, to regard some parts as of *negative* density ; but we must recollect that we are dealing with the *difference* between a disturbed and an undisturbed system, which may be positive or negative, though no real mass can be negative.

Let us for an instant conceive only one of these surfaces to exist, and let us transfer the origin to it. Then the law of thickness is

$$\zeta = mcA \sin mx \dots\dots\dots\dots\dots\dots\dots\dots \text{(83)},$$

and we know that the normal component of attraction at the surface is the same as if the thickness had been uniform throughout, so that

$$\frac{dV}{dz} = -2\pi k\zeta,$$

on the positive side of the surface.

Also, the solution of the equation

$$\frac{d^2V}{dx^2} + \frac{d^2V}{dz^2} = 0,$$

consists of a series of terms of the form $C\epsilon^{iz} \sin ix$.

Of these the only one with which we have to do is that in which $i = -m$. Applying the condition as to the normal force at the surface, we get

$$V = 2\pi kc\epsilon^{-mz}A \sin mx \dots\dots\dots\dots\dots\dots \text{(87)},$$

for the potential on the positive side of the surface, and

$$V = 2\pi kc\epsilon^{mz}A \sin mx \dots\dots\dots\dots\dots\dots\text{(88)},$$

on the negative side.

Calculating the potentials of a pair of such surfaces at distances $+c$ and $-c$ from the plane of xy, and calling V' the sum of their potentials, we have for the space between these planes

$$V_1' = 2\pi kcA \sin mx\epsilon^{-mc}\left(\epsilon^{mz} + \epsilon^{-mz}\right)$$

beyond them

$$V_2' = 2\pi kcA \sin mx\epsilon^{\mp mz}\left(\epsilon^{mc} + \epsilon^{-mc}\right)$$

.... (89);

the upper or lower sign of the index being taken according as z is positive or negative.

These potentials must be added to those formerly obtained, to get the potential at any point after displacement.

We have next to calculate the pressure of the fluid at any point, on the supposition that the imaginary planes protect each slice of the fluid from the pressure of the adjacent slices, so that it is in equilibrium under the action of the forces of attraction, and the pressure of these planes on each side. Now in a fluid of density k, in equilibrium under forces whose potential is V, we have always—

$$\frac{dp}{dV} = k;$$

so that if we know that the value of p is p_0 where that of V is V_0, then at any other point

$$p = p_0 + k\left(V - V_0\right).$$

Now, at the free surface of the fluid, $p = 0$, and the distance from the free surface of the disturbed fluid to the plane of the original surface is ζ, a small quantity. The attraction which acts on this stratum of fluid is, in the first place, that of the undisturbed stratum, and this is equal to $4\pi kc$, towards that stratum. The pressure due to this cause at the level of the original surface will be $4\pi k^2 c\zeta$, and the pressure arising from the attractive forces due to the displacements upon this thin layer of fluid, will be small quantities of the second order, which we neglect. We thus find the pressure when $z = c$ to be,

$$p_0 = 4\pi k^2 c^2 mA \sin mx.$$

The potential of the undisturbed mass when $z = c$ is

$$V_0 = -2\pi kc^2,$$

and the potential of the disturbance itself for the same value of z, is

$$V_0' = 2\pi kcA \sin mx\left(1 + \epsilon^{-2mc}\right).$$

So that we find the general value of p at any other point to be

$$p = 2\pi k^2 \left(c^2 - z^2\right) + 2\pi k^2 cA \sin mx \left\{2cm - 1 - \epsilon^{-2mc} + \epsilon^{mc} \left(\epsilon^{mz} + \epsilon^{-mz}\right)\right\} \dots (90).$$

This expression gives the pressure of the fluid at any point, as depending on the state of constraint produced by the displacement of the imaginary planes. The accelerating effect of these pressures on any particle, if it were allowed to move parallel to x, instead of being confined by the planes, would be

$$= \frac{1}{k} \frac{dp}{dx}.$$

The accelerating effect of the attractions in the same direction is

$$\frac{dV}{dx},$$

so that the whole acceleration parallel to x is

$$X = -2\pi kmcA \cos mx \left(2mc - \epsilon^{-2mc} - 1\right) \dots\dots\dots\dots (91).$$

It is to be observed, that this quantity is independent of z, so that every particle in the slice, by the combined effect of pressure and attraction, is urged with the same force, and, if the imaginary planes were removed, each slice would move parallel to itself without distortion, as long as the absolute displacements remained small. We have now to consider the direction of the resultant force X, and its changes of magnitude.

We must remember that the original displacement is $A \cos mx$, if therefore $(2mc - \epsilon^{-2mc} - 1)$ be positive, X will be opposed to the displacement, and the equilibrium will be stable, whereas if that quantity be negative, X will act along with the displacement and increase it, and so constitute an unstable condition.

It may be seen that large values of mc give positive results and small ones negative. The sign changes when

$$2mc = 1 \cdot 147 \dots\dots\dots\dots\dots\dots\dots\dots\dots\dots\dots\dots (92),$$

which corresponds to a wave-length

$$\lambda = 2c \frac{2\pi}{1 \cdot 147} = 2c \left(5 \cdot 471\right) \dots\dots\dots\dots\dots\dots\dots (93).$$

The length of the complete wave in the critical case is $5 \cdot 471$ times the thickness of the stratum. Waves shorter than this are stable, longer waves are unstable.

The quantity $\qquad\qquad$ $2mc\left(2mc - \epsilon^{-2mc} - 1\right),$

has a minimum when $\qquad\qquad$ $2mc = \cdot607$ ....(94),

and the wave-length is $10\cdot353$ times the thickness of the stratum.

In this case $\qquad\qquad$ $2mc\left(2mc - \epsilon^{-2mc} - 1\right) = -\cdot509$ ....(95),

and $\qquad\qquad\qquad$ $X = \cdot509\pi kA \cos mx$ ....(96).

24. Let us now conceive that the stratum of fluid, instead of being infinite in extent, is limited in breadth to about 100 times the thickness. The pressures and attractions will not be much altered by this removal of a distant part of the stratum. Let us also suppose that this thin but broad strip is bent round in its own plane into a circular ring whose radius is more than ten times the breadth of the strip, and that the waves, instead of being exactly parallel to each other, have their ridges in the direction of radii of the ring. We shall then have transformed our stratum into one of Saturn's Rings, if we suppose those rings to be liquid, and that a considerable breadth of the ring has the same angular velocity.

Let us now investigate the conditions of stability by putting

$$x = -2\pi kmc\left(2mc - \epsilon^{-2mc} - 1\right)$$

into the equation for n. We know that x must lie between 0 and $\dfrac{\omega^2}{13\cdot9}$ to ensure stability. Now the greatest value of x in the fluid stratum is $\cdot509\pi k$. Taking Laplace's ratio of the diameter of the ring to that of the planet, this gives $42\cdot5$ as the minimum value of the density of the planet divided by that of the fluid of the ring.

Now Laplace has shewn that any value of this ratio greater than $1\cdot3$ is inconsistent with the rotation of any considerable breadth of the fluid at the same angular velocity, so that our hypothesis of a broad ring with uniform velocity is untenable.

But the stability of such a ring is impossible for another reason, namely, that for waves in which $2mc > 1\cdot147$, x is negative, and the ring will be destroyed by these short waves in the manner described at page (333).

When the fluid ring is treated, not as a broad strip, but as a filament of circular or elliptic section, the mathematical difficulties are very much increased,

but it may be shown that in this case also there will be a maximum value
of x, which will require the density of the planet to be several times that of
the ring, and that in all cases short waves will give rise to negative values
of x, inconsistent with the stability of the ring.

It appears, therefore, that a ring composed of a continuous liquid mass
cannot revolve about a central body without being broken up, but that the
parts of such a broken ring may, under certain conditions, form a permanent
ring of satellites.

On the Mutual Perturbations of Two Rings.

25.   We shall assume that the difference of the mean radii of the rings
is small compared with the radii themselves, but large compared with the
distance of consecutive satellites of the same ring.   We shall also assume that
each ring separately satisfies the conditions of stability.

We have seen that the effect of a disturbing force on a ring is to produce
a series of waves whose number and period correspond with those of the disturbing force which produces them, so that we have only to calculate the
coefficient belonging to the wave from that of the disturbing force.

Hence in investigating the simultaneous motions of two rings, we may
assume that the mutually disturbing waves travel with the same *absolute*
angular velocity, and that a maximum in one corresponds either to a maximum
or a minimum of the other, according as the coefficients have the same or
opposite signs.

Since the motions of the particles of each ring are affected by the disturbance
of the other ring, as well as of that to which they belong, the equations of
motion of the two rings will be involved in each other, and the final equation
for determining the wave-velocity will have eight roots instead of four.   But as
each of the rings has four *free* waves, we may suppose these to originate *forced*
waves in the other ring, so that we may consider the eight waves of each ring
as consisting of four free waves and four forced ones.

In strictness, however, the wave-velocity of the "free" waves will be
affected by the existence of the forced waves which they produce in the other
ring, so that none of the waves are really "free" in either ring independently,
though the whole motion of the system of two rings as a whole is free.

We shall find, however, that it is best to consider the waves first as free, and then to determine the reaction of the other ring upon them, which is such as to alter the wave-velocity of both, as we shall see.

The forces due to the second ring may be separated into three parts.

1st. The constant attraction when both rings are at rest.

2nd. The variation of the attraction on the first ring, due to its own disturbances.

3rd. The variation of the attraction due to the disturbances of the second ring.

The first of these affects only the angular velocity. The second affects the waves of each ring independently, and the mutual action of the waves depends entirely on the third class of forces.

26. *To determine the attractions between two rings.*

Let R and a be the mass and radius of the exterior ring, R' and a' those of the interior, and let all quantities belonging to the interior ring be marked with accented letters. (Fig. 5.)

1st. *Attraction between the rings when at rest.*

Since the rings are at a distance small compared with their radii, we may calculate the attraction on a particle of the first ring as if the second were an infinite straight line at distance $a' - a$ from the first.

The mass of unit of length of the second ring is $\dfrac{R'}{2\pi a'}$, and the accelerating effect of the attraction of such a filament on an element of the first ring is

$$\frac{R'}{\pi a'\,(a - a')} \text{ inwards } \dots\dots\dots\dots\dots\dots\dots\dots\dots(97).$$

The attraction of the first ring on the second may be found by transposing accented and unaccented letters.

In consequence of these forces, the outer ring will revolve faster, and the inner ring slower than would otherwise be the case. These forces enter into the *constant terms* of the equations of motion, and may be included in the value of K.

2nd.   *Variation due to disturbance of first ring.*

If we put $a(1+\rho)$ for a in the last expression, we get the attraction when the first ring is displaced. The part depending on ρ is

$$-\frac{R'a}{\pi a'(a-a')^2}\,\rho \text{ inwards} \dots\dots\dots\dots\dots\dots(98).$$

This is the only variation of force arising from the displacement of the first ring. It affects the value of L in the equations of motion.

3rd.   *Variation due to waves in the second ring.*

On account of the waves, the second ring varies in distance from the first, and also in mass of unit of length, and each of these alterations produces variations both in the radial and tangential force, so that there are four things to be calculated :

1st.   Radial force due to radial displacement.

2nd.   Radial force due to tangential displacement.

3rd.   Tangential force due to radial displacement.

4th.   Tangential force due to tangential displacement.

1st.   Put $a'(1+\rho')$ for a', and we get the term in ρ'

$$\frac{R'}{\pi a'}\frac{(2a'-a)}{(a'-a)^3}\,\rho' \text{ inwards} = \lambda'\rho', \text{ say} \dots\dots\dots\dots (99).$$

2nd.   By the tangential displacement of the second ring the section is reduced in the proportion of 1 to $1-\dfrac{d\sigma'}{ds'}$, and therefore there is an alteration of the radial force equal to

$$-\frac{R'}{\pi a'(a-a')}\frac{d\sigma'}{ds'} \text{ inwards} = -\mu'\frac{d\sigma'}{ds'} \text{ say} \dots\dots\dots\dots (100).$$

3rd.   By the radial displacement of the second ring the direction of the filament near the part in question is altered, so that the attraction is no longer radial but forwards, and the tangential part of the force is

$$\frac{R'}{\pi a'(a-a')}\frac{d\rho'}{ds'} = +\mu'\frac{d\rho'}{ds'} \text{ forwards} \dots\dots\dots\dots (101).$$

4th. By the tangential displacement of the second ring a tangential force arises, depending on the relation between the length of the waves and the distance between the rings.

If we make $m\,\dfrac{a-a'}{a'}=p$, and $m\displaystyle\int_{-\infty}^{+\infty}\dfrac{x\sin px}{(1+x^2)^{\frac{3}{2}}}\,dx=\Pi,$

the tangential force is $\dfrac{R'}{\pi a'\,(a-a')^2}\,\Pi\sigma'=\nu'\sigma' \ \ldots\ldots\ldots\ldots\ldots\ (102).$

We may now write down the values of λ, μ, and ν by transposing accented and unaccented letters.

$$\lambda=\frac{R}{\pi a}\frac{(2a-a')}{(a-a')^2}\,;\ \mu=\frac{R}{\pi a\,(a'-a)}\,;\ \nu=\frac{R}{\pi a'\,(a-a')^2}\,\Pi \ \ldots\ldots\ (103).$$

Comparing these values with those of λ', μ', and ν', it will be seen that the following relations are approximately true when a is nearly equal to a':

$$\frac{\lambda'}{\lambda}=-\frac{\mu'}{\mu}=\frac{\nu'}{\nu}=\frac{R'a}{Ra'} \ \ldots\ldots\ldots\ldots\ldots\ldots\ (104).$$

27. To form the equations of motion.

*The original equations were

$$\omega^2+\omega^2\rho+2\omega\,\frac{d\sigma}{dt}-\frac{d^2\rho}{dt^2}=P=S+K-(2S-L)\,A\rho-MB\rho+\lambda'\rho'-\mu'\,\frac{d\sigma'}{ds'},$$

$$2\omega\,\frac{d\rho}{dt}+\frac{d^2\sigma}{dt^2}=Q=MA\sigma+NB\sigma+\mu'\,\frac{d\rho'}{ds'}+\nu'\sigma'.$$

Putting
$$\rho=A\cos{(ms+nt)},\quad \sigma=B\sin{(ms+nt)},$$
$$\rho'=A'\cos{(ms+nt)},\quad \sigma'=B'\sin{(ms+nt)},$$
then
$$\omega^2=S+K$$
$$\left.\begin{array}{l}(\omega^2+2S+n^2-L)\,A+(2\omega n+M)\,B-\lambda'A'+\mu'mB'=0\\ (2\omega n+M)\,A+(n^2+N)\,B-\mu'mA'+\nu'B'=0\end{array}\right\}\ldots\ldots(105).$$

The corresponding equations for the second ring may be found by transposing accented and unaccented letters. We should then have four equations to determine the ratios of A, B, A', B', and a resultant equation of the eighth degree to determine n. But we may make use of a more convenient method, since λ', μ', and ν' are small. Eliminating B we find

$$\left.\begin{array}{l}An^4-A\,(\omega^2+2K+L-N)\,n^2-4A\omega Mn+AN\,(3\omega^2)\\ (-\lambda'A'+\mu'mB')\,n^2+(\mu'mA'-\nu'B')\,2\omega n\end{array}\right\}=0\ \ldots\ldots\ldots(106).$$

* [The analysis in this article is somewhat unsatisfactory, the equations of motion employed being those which were applicable in the case of a ring of radius unity. ED.]

Putting
$$B = \beta A, \quad A' = xA, \quad B' = \beta' A' = \beta' xA,$$

we have
$$\left.\begin{array}{l} n^4 - \{\omega^2(+2K) + L - N\}\, n^2 - 4\omega Mn + 3\omega^2 N \\ \quad + (-\lambda' + \mu' m\beta')\, n^2 x + (\mu' m - \nu'\beta')\, 2\omega nx \end{array}\right\} = U = 0 \dots\dots(107).$$

$$\frac{dU}{dn} = 4n^3 - 2\omega^2 n + \&c. \dots\dots\dots\dots\dots\dots\dots\dots\dots\dots\dots (108),$$

$$\frac{dU}{dx} = -\lambda' n^2 + \mu' m\beta' n^2 + 2\mu' m\omega n - 2\nu'\beta'\omega n \dots\dots\dots\dots(109),$$

whence
$$\frac{dn}{dx} = \frac{\lambda' n - \mu' m\beta' n - 2\mu' m\omega + 2\nu'\beta'\omega}{4n^2 - 2\omega^2} \dots\dots\dots\dots(110).$$

28. If we were to solve the equation for n, leaving out the terms involving x, we should find the wave-velocities of the four free waves of the first ring, supposing the second ring to be prevented from being disturbed. But in reality the waves in the first ring produce a disturbance in the second, and these in turn react upon the first ring, so that the wave-velocity is somewhat different from that which it would be in the supposed case. Now if x be the ratio of the radial amplitude of displacement in the second ring to that in the first, and if $\bar{n}$ be a value of n supposing $x = 0$, then by Maclaurin's theorem,

$$n = +\bar{n} + \frac{dn}{dx}\, x \dots\dots\dots\dots\dots\dots\dots\dots\dots(111).$$

The wave-velocity relative to the ring is $-\dfrac{n}{m}$, and the absolute angular velocity of the wave in space is

$$\varpi = \omega - \frac{n}{m} = \omega - \frac{\bar{n}}{m} - \frac{1}{m}\frac{dn}{dx}\, x \dots\dots\dots\dots\dots(112),$$

$$= +p - qx \dots\dots\dots\dots\dots\dots\dots (113),$$

where $p = \omega - \dfrac{\bar{n}}{m}$, and $q = \dfrac{1}{m}\dfrac{dn}{dx}$.

Similarly in the second ring we should have

$$\varpi' = p' - q'\frac{1}{x} \dots\dots\dots\dots\dots\dots\dots\dots(114);$$

and since the corresponding waves in the two rings must have the same absolute angular velocity,

$$\varpi = \varpi', \text{ or } p - qx = p' - q'\frac{1}{x} \dots\dots\dots\dots\dots (115).$$

This is a quadratic equation in x, the roots of which are real when

$$(p-p')^2 + 4qq'$$

is positive. When this condition is not fulfilled, the roots are impossible, and the general solution of the equations of motion will contain exponential factors, indicating destructive oscillations in the rings.

Since q and q' are small quantities, the solution is always real whenever p and p' are considerably different. The absolute angular velocities of the two pairs of reacting waves, are then nearly

$$p + \frac{qq'}{p-p'}, \text{ and } p' - \frac{qq'}{p-p'},$$

instead of p and p', as they would have been if there had been no reaction of the forced wave upon the free wave which produces it.

When p and p' are equal or nearly equal, the character of the solution will depend on the sign of qq'. We must therefore determine the signs of q and q' in such cases.

Putting $\beta' = \dfrac{2\omega'}{n'}$, we may write the values of q and q'

$$\left.
\begin{aligned}
q &= \frac{n}{m} \cdot \frac{\lambda' + 2\mu'm \left(\dfrac{\omega'}{n'} - \dfrac{\omega}{n}\right) - 4\nu'\dfrac{\omega'}{n'}\dfrac{\omega}{n}}{4n^2 - 2\omega^2} \\[2ex]
q' &= \frac{n'}{m'} \cdot \frac{\lambda + 2\mu m \left(\dfrac{\omega}{n} - \dfrac{\omega'}{n'}\right) - 4\nu\dfrac{\omega'}{n'}\dfrac{\omega}{n}}{4n'^2 - 2\omega'^2}
\end{aligned}
\right\} \dots\dots\dots\dots (116).$$

Referring to the values of the disturbing forces, we find that

$$\frac{\lambda'}{\lambda} = -\frac{\mu'}{\mu} = \frac{\nu'}{\nu} = \frac{R'a}{Ra'}.$$

Hence
$$\frac{q}{q'} = \frac{n}{n'} \frac{4n'^2 - 2\omega'^2}{4n^2 - 2\omega^2} \cdot \frac{R'a}{Ra'} \dots\dots\dots\dots\dots (117).$$

Since qq' is of the same sign as $\dfrac{q}{q'}$, we have only to determine whether $2n - \dfrac{\omega^2}{n}$, and $2n' - \dfrac{\omega'^2}{n}$, are of the same or of different signs. If these quantities are of the same sign, qq' is positive, if of different signs, qq' is negative.

Now there are four values of n, which give four corresponding values of $2n - \dfrac{\omega^2}{n}$:

$$n_1 = -\omega + \&c., \qquad\qquad 2n_1 - \frac{\omega^2}{n_1} \text{ is negative,}$$

$$n_2 = -\text{a small quantity, } 2n_2 - \frac{\omega^2}{n_2} \text{ is positive,}$$

$$n_3 = +\text{a small quantity, } 2n_3 - \frac{\omega^2}{n_3} \text{ is negative,}$$

$$n_4 = \omega - \&c., \qquad\qquad 2n_4 - \frac{\omega^2}{n_4} \text{ is positive.}$$

The quantity with which we have to do is therefore positive for the even orders of waves and negative for the odd ones, and the corresponding quantity in the other ring obeys the same law. Hence when the waves which act upon each other are either both of even or both of odd names, qq' will be positive, but when one belongs to an even series, and the other to an odd series, qq' is negative.

29. The values of p and p' are, roughly,

$$\left.\begin{aligned}
p_1 &= \omega + \frac{\omega}{m} - \&c., \ p_2 = \omega + \&c., \ p_3 = \omega - \&c., \ p_4 = \omega - \frac{\omega}{m} + \&c. \\
p_1' &= \omega' + \frac{\omega'}{m} - \&c., \ p_2' = \omega' + \&c., \ p_3' = \omega' - \&c., \ p_4' = \omega' - \frac{\omega'}{m} + \&c.
\end{aligned}\right\} \ \ldots\ldots\ldots (118).$$

ω' is greater than ω, so that p_1' is the greatest, and p_4 the least of these values, and of those of the same order, the accented is greater than the unaccented. The following cases of equality are therefore possible under suitable circumstances:

$$
\begin{array}{ll}
p_1 = p_3', & p_1 = p_2', \\
p_2 = p_4', & p_1 = p_4', \\
p_4 = p_4' \ (\text{when } m = 1), & p_2 = p_3', \\
& p_3 = p_4',
\end{array}
$$

In the cases in the first column qq' will be positive, in those in the second column qq' will be negative.

30. Now each of the four values of p is a function of m, the number of undulations in the ring, and of a the radius of the ring, varying nearly as $a^{-\frac{3}{2}}$. Hence m being given, we may alter the radius of the ring till any one of the four values of p becomes equal to a given quantity, say a given value of p', so that if an indefinite number of rings coexisted, so as to form a sheet of rings, it would be always possible to discover instances of the equality of p and p' among them. If such a case of equality belongs to the first column given above, two constant waves will arise in both rings, one travelling a little faster, and the other a little slower than the free waves. If the case belongs to the second column, two waves will also arise in each ring, but the one pair will gradually die away, and the other pair will increase in amplitude indefinitely, the one wave strengthening the other till at last both rings are thrown into confusion.

The only way in which such an occurrence can be avoided is by placing the rings at such a distance that no value of m shall give coincident values of p and p'. For instance, if $\omega' > 2\omega$, but $\omega' < 3\omega$, no such coincidence is possible. For p_1 is always less than p_2', it is greater than p_4 when $m = 1$ or 2, and less than p_4 when m is 3 or a greater number. There are of course an infinite number of ways in which this noncoincidence might be secured, but it is plain that if a number of concentric rings were placed at small intervals from each other, such coincidences must occur accurately or approximately between some pairs of rings, and if the value of $(p - p')^2$ is brought lower than $-4qq'$, there will be destructive interference.

This investigation is applicable to any number of concentric rings, for, by the principle of superposition of small displacements, the reciprocal actions of any pair of rings are independent of all the rest.

31. *On the effect of long-continued disturbances on a system of rings.*

The result of our previous investigations has been to point out several ways in which disturbances may accumulate till collisions of the different particles of the rings take place. After such a collision the particles will still continue to revolve about the planet, but there will be a loss of energy in the system during the collision which can never be restored. Such collisions however will not affect what is called the Angular Momentum of the system about the planet, which will therefore remain constant.

Let M be the mass of the system of rings, and δm that of one ring whose radius is r, and angular velocity $\omega = S^{\frac{1}{2}} r^{-\frac{3}{2}}$. The angular momentum of the ring is

$$\omega r^2 \delta m = S^{\frac{1}{2}} r^{\frac{1}{2}} \delta m,$$

half its *vis viva* is
$$\tfrac{1}{2}\omega^2 r^2 \delta m = \tfrac{1}{2} S r^{-1}\,\delta m.$$

The potential energy due to Saturn's attraction on the ring is

$$- S r^{-1} \delta m.$$

The angular momentum of the whole system is invariable, and is

$$S^{\frac{1}{2}} \Sigma \left(r^{\frac{1}{2}} \delta m \right) = A \quad\ldots\ldots\ldots\ldots\ldots\ldots\ldots (119).$$

The whole energy of the system is the sum of half the *vis viva* and the potential energy, and is

$$- \tfrac{1}{2} S \Sigma \left(r^{-1}\,\delta m \right) = E \quad\ldots\ldots\ldots\ldots\ldots\ldots (120).$$

A is invariable, while E necessarily diminishes. We shall find that as E diminishes, the distribution of the rings must be altered, some of the outer rings moving outwards, while the inner rings move inwards, so as either to spread out the whole system more, both on the outer and on the inner edge of the system, or, without affecting the extreme rings, to diminish the density or number of the rings at the mean distance, and increase it at or near the inner and outer edges.

Let us put $x = r^{\frac{1}{2}}$, then $A = S^{\frac{1}{2}} \Sigma \left(x\,dm \right)$ is constant.

Now let
$$x_1 = \frac{\Sigma\left(x\,dm\right)}{\Sigma\left(dm\right)},$$

and
$$x = x_1 + x',$$

then we may write

$$- \frac{2E}{S} = \Sigma \left(r^{-1}\,\delta m \right) = \Sigma \left(x^{-2}\,dm \right),$$

$$= \Sigma dm \left(x_1^{-2} - 2\,\frac{x'}{x_1^3} + 3\,\frac{x'^2}{x_1^4} - \&\text{c.} \right),$$

$$= \frac{1}{x_1^2}\,\Sigma \left(dm \right) - \frac{2}{x_1^3}\,\Sigma \left(x'\,dm \right) + \frac{3}{x_1^4}\,\Sigma \left(x'^2 \delta m \right) - \&\text{c.}\ldots\ldots(121).$$

Now $\Sigma\,(dm) = M$ a constant, $\Sigma\,(x'dm) = 0$, and $\Sigma\,(x'^2\delta m)$ is a quantity which increases when the rings are spread out from the mean distance either way, x' being subject only to the restriction $\Sigma\,(x'dm) = 0$. But $\Sigma\,(x'^2dm)$ may increase without the extreme values of x' being increased, provided some other values be increased.

32. In fact, if we consider the very innermost particle as moving in an ellipse, and at the further apse of its orbit encountering another particle belonging to a larger orbit, we know that the second particle, when at the same distance from the planet, moves the faster. The result is, that the interior satellite will receive a forward impulse at its further apse, and will move in a larger and less eccentric orbit than before. In the same way one of the outermost particles may receive a backward impulse at its nearer apse, and so be made to move in a smaller and less eccentric orbit than before. When we come to deal with collisions among bodies of unknown number, size, and shape, we can no longer trace the mathematical laws of their motion with any distinctness. All we can now do is to collect the results of our investigations and to make the best use we can of them in forming an opinion as to the constitution of the actual rings of Saturn which are still in existence and apparently in steady motion, whatever catastrophes may be indicated by the various theories we have attempted.

33. *To find the Loss of Energy due to internal friction in a broad Fluid Ring, the parts of which revolve about the Planet, each with the velocity of a satellite at the same distance.*

Conceive a fluid, the particles of which move parallel to the axis of x with a velocity u, u being a function of z, then there will be a tangential pressure on a plane parallel to xy

$$= \mu \frac{du}{dz}\ \text{on unit of area}$$

due to the relative sliding of the parts of the fluid over each other.

In the case of the ring we have

$$\omega = S^{\frac{1}{2}}\, r^{-\frac{3}{2}}.$$

The absolute velocity of any particle is ωr. That of a particle at distance $(r + \delta r)$ is

$$\omega r + \frac{d}{dr}\,(\omega r)\,\delta r.$$

If the angular velocity had been uniform, there would have been no sliding, and the velocity would have been

$$\omega r + \omega \delta r.$$

The sliding is therefore

$$r \frac{d\omega}{dr} \delta r,$$

and the friction on unit of area perpendicular to r is $\mu r \dfrac{d\omega}{dr}$.

The loss of Energy, per unit of area, is the product of the sliding by the friction,

or, $\mu r^2 \overline{\dfrac{d\omega}{dr}}\Big|^2 \delta r$ in unit of time.

The loss of Energy in a part of the Ring whose radius is r, breadth δr, and thickness c, is

$$2\pi r^3 c\mu \, \overline{\frac{d\omega}{dr}}\Big|^2 \delta r.$$

In the case before us it is $\tfrac{9}{2}\pi\mu Scr^{-2}\,\delta r.$

If the thickness of the ring is uniform between $r = a$ and $r = b$, the whole loss of Energy is

$$\tfrac{9}{2}\pi\mu Sc \left(\frac{1}{b} - \frac{1}{a}\right),$$

in unit of time.

Now half the *vis viva* of an elementary ring is

$$\pi\rho c r \delta r \; r^2\omega^2 = \pi\rho c S \delta r,$$

and this between the limits $r = a$ and $r = b$ gives

$$\pi\rho c S \,(a - b).$$

The potential due to the attraction of S is twice this quantity with the sign changed, so that

$$E = -\pi\rho c S \,(a - b),$$

and $\dfrac{dE}{dt} = \tfrac{9}{2}\pi\mu S \left(\dfrac{1}{b} - \dfrac{1}{a}\right),$

$$\frac{1}{E} \frac{dE}{dt} = -\tfrac{9}{2} \frac{\mu}{\rho} \frac{1}{ab}.$$

Now Professor Stokes finds $\sqrt{\dfrac{\mu}{\rho}} = 0.0564$ for water,

and $\qquad\qquad\qquad\qquad\qquad = 0.116$ for air,

taking the unit of space one English inch, and the unit of time one second. We may take $a = 88,209$ miles, and $b = 77,636$ for the ring A ; and $a = 75,845$, and $b = 58,660$ for the ring B. We may also take one year as the unit of time. The quantity representing the ratio of the loss of energy in a year to the whole energy is

$$\frac{1}{E}\frac{dE}{dt} = \frac{1}{60,880,000,000,000} \text{ for the ring } A,$$

$$\text{and } \frac{1}{39,540,000,000,000} \text{ for the ring } B,$$

showing that the effect of internal friction in a ring of water moving with steady motion is inappreciably small. It cannot be from this cause therefore that any decay can take place in the motion of the ring, provided that no waves arise to disturb the motion.

Recapitulation of the Theory of the Motion of a Rigid Ring.

The position of the ring relative to Saturn at any given instant is defined by three variable quantities.

1st. The distance between the centre of gravity of Saturn and the centre of gravity of the ring. This distance we denote by r.

2nd. The angle which the line r makes with a fixed line in the plane of the motion of the ring. This angle is called θ.

3rd. The angle between the line r and a line fixed with respect to the ring so that it coincides with r when the ring is in its mean position. This is the angle ϕ.

The values of these three quantities determine the position of the ring so far as its motion in its own plane is concerned. They may be referred to as the *radius vector, longitude,* and *angle of libration* of the ring.

The forces which act between the ring and the planet depend entirely upon their relative positions. The method adopted above consists in determining the *potential* (V) of the ring at the centre of the planet in terms of r and ϕ. Then the *work done* by any displacement of the system is measured by the change of VS during that displacement. The attraction between the centre of gravity of the Ring and that of the planet is $-S\dfrac{dV}{dr}$, and the moment of the couple tending to turn the ring about its centre of gravity is $S\dfrac{dV}{d\phi}$.

It is proved in Problem V, that if a be the radius of a circular ring, $r_0 = af$ the distance of its centre of gravity from the centre of the circle, and R the mass of the ring, then, at the centre of the ring, $\dfrac{dV}{dr} = -\dfrac{R}{a^2}f,\ \dfrac{dV}{d\phi} = 0$.

It also appears that $\dfrac{d^2V}{dr^2} = \tfrac{1}{2}\dfrac{R}{a^3}(1+g)$, which is positive when $g > -1$,

and that $\dfrac{d^2V}{d\phi^2} = \tfrac{1}{2}\dfrac{R}{a}f^2(3-g)$, which is positive when $g < 3$.

If $\dfrac{d^2V}{dr^2}$ is positive, then the attraction between the centres decreases as the distance increases, so that, if the two centres were kept at rest at a given distance by a constant force, the equilibrium would be unstable. If $\dfrac{d^2V}{d\phi^2}$ is positive, then the forces tend to increase the angle of libration, in whichever direction the libration takes place, so that if the ring were fixed by an axis through its centre of gravity, its equilibrium round that axis would be unstable.

In the case of the uniform ring with a heavy particle on its circumference whose weight = ·82 of the whole, the direction of the whole attractive force of the ring near the centre will pass through a point lying in the same radius as the centre of gravity, but at a distance from the centre = $\tfrac{8}{9}a$. (Fig. 6.)

If we call this point O, the line SO will indicate the direction and position of the force acting on the ring, which we may call F.

It is evident that the force F, acting on the ring in the line OS, will tend to turn it round its centre of gravity R and to increase the angle of libration KRO. The direct action of this force can never reduce the angle of libration to zero again. To understand the indirect action of the force, we must recollect that the centre of gravity (R) of the ring is revolving about Saturn in the direction of the arrows, and that the ring is revolving about its centre of gravity with nearly the same velocity. If the angular velocity of the centre of gravity about Saturn were always equal to the rotatory velocity of the ring, there would be no libration.

Now suppose that the angle of rotation of the ring is in advance of the longitude of its centre of gravity, so that the line RO has got in advance of SRK by the angle of libration KRO. The attraction between the planet and the ring is a force F acting in SO. We resolve this force into a couple, whose moment is $F \cdot RN$, and a force F acting through R the centre of gravity of the ring.

The couple affects the rotation of the ring, but not the position of its centre of gravity, and the force RF acts on the centre of gravity without affecting the rotation.

Now the couple, in the case represented in the figure, acts in the positive direction, so as to *increase* the angular velocity of the ring, which was already greater than the velocity of revolution of R about S, so that the angle of libration would increase, and never be reduced to zero.

The force RF does not act in the direction of S, but behind it, so that it becomes a retarding force acting upon the centre of gravity of the ring. Now the effect of a retarding force is to cause the distance of the revolving body to decrease and the angular velocity to increase, so that a retarding force increases the angular velocity of R about S.

The effect of the attraction along SO in the case of the figure is, first, to increase the rate of rotation of the ring round R, and secondly, to increase the angular velocity of R about S. If the second effect is greater than the first, then, although the line RO increases its angular velocity, SR will increase its angular velocity more, and will overtake RO, and restore the ring to its original position, so that SRO will be made a straight line as at first. If this accelerating effect is not greater than the acceleration of rotation about R due to the couple, then no compensation will take place, and the motion will be essentially unstable.

If in the figure we had drawn ϕ negative instead of positive, then the couple would have been negative, the tangential force on R accelerative, r would have increased, and in the cases of stability the retardation of θ would be greater than that of $(\theta + \phi)$, and the normal position would be restored, as before.

The object of the investigation is to find the conditions under which this compensation is possible.

It is evident that when SRO becomes straight, there is still a difference of angular velocities between the rotation of the ring and the revolution of the centre of gravity, so that there will be an oscillation on the other side, and the motion will proceed by alternate oscillations without limit.

If we begin with r at its mean value, and ϕ negative, then the rotation of the ring will be retarded, r will be increased, the revolution of r will be more retarded, and thus ϕ will be reduced to zero. The next part of the motion will reduce r to its mean value, and bring ϕ to its greatest positive value. Then r will diminish to its least value, and ϕ will vanish. Lastly r will return to the mean value, and ϕ to the greatest negative value.

It appears from the calculations, that there are, in general, two different ways in which this kind of motion may take place, and that these may have different periods, phases, and amplitudes. The mental exertion required in following out the results of a combined motion of this kind, with all the variations of force and velocity during a complete cycle, would be very great in proportion to the additional knowledge we should derive from the exercise.

The result of this theory of a rigid ring shows not only that a perfectly uniform ring cannot revolve permanently about the planet, but that the irregularity of a permanently revolving ring must be a very observable quantity, the distance between the centre of the ring and the centre of gravity being between ·8158 and ·8279 of the radius. As there is no appearance about the rings justifying a belief in so great an irregularity, the theory of the solidity of the rings becomes very improbable.

When we come to consider the additional difficulty of the tendency of the fluid or loose parts of the ring to accumulate at the thicker parts, and thus to destroy that nice adjustment of the load on which stability depends, we have another powerful argument against solidity.

And when we consider the immense size of the rings, and their comparative thinness, the absurdity of treating them as rigid bodies becomes self-evident. An iron ring of such a size would be not only plastic but semifluid under the forces which it would experience, and we have no reason to believe these rings to be artificially strengthened with any material unknown on this earth.

Recapitulation of the Theory of a Ring of equal Satellites.

In attempting to conceive of the disturbed motion of a ring of unconnected satellites, we have, in the first place, to devise a method of identifying each satellite at any given time, and in the second place, to express the motion of every satellite under the same general formula, in order that the mathematical methods may embrace the whole system of bodies at once.

By conceiving the ring of satellites arranged regularly in a circle, we may easily identify any satellite, by stating the angular distance between it and a known satellite when so arranged. If the motion of the ring were undisturbed, this angle would remain unchanged during the motion, but, in reality, the satellite has its position altered in three ways: 1st, it may be further from or nearer to Saturn; 2ndly, it may be in advance or in the rear of the position it would have had if undisturbed; 3rdly, it may be on one side or other of the mean plane of the ring. Each of these displacements may vary in any way whatever as we pass from one satellite to another, so that it is impossible to assign beforehand the place of any satellite by knowing the places of the rest. § 2.

The formula, therefore, by which we are enabled to predict the place of every satellite at any given time, must be such as to allow the initial position of every satellite to be independent of the rest, and must express all future positions of that satellite by inserting the corresponding value of the quantity denoting time, and those of every other satellite by inserting the value of the angular distance of the given satellite from the point of reference. The three displacements of the satellite will therefore be functions of two variables—the angular position of the satellite, and the time. When the time alone is made to vary, we trace the complete motion of a single satellite; and when the time is made constant, and the angle is made to vary, we trace the form of the ring at a given time.

It is evident that the form of this function, in so far as it indicates the state of the whole ring at a given instant, must be wholly arbitrary, for the form of the ring and its motion at starting are limited only by the condition that the irregularities must be small. We have, however, the means of breaking up any function, however complicated, into a series of simple functions, so that the value of the function between certain limits may be accurately expressed as the sum of a series of sines and cosines of multiples of the variable. This method, due to Fourier, is peculiarly applicable to the case of a ring returning into itself, for the value of Fourier's series is necessarily periodic. We now regard the form of the disturbed ring at any instant as the result of the superposition of a number of separate disturbances, each of which is of the nature of a series of equal waves regularly arranged round the ring. Each of these elementary disturbances is characterised by the number of undulations in it, by their amplitude, and by the position of the first maximum in the ring. § 3.

When we know the form of each elementary disturbance, we may calculate the attraction of the disturbed ring on any given particle in terms of the constants belonging to that disturbance, so that as the actual displacement is the resultant of the elementary displacements, the actual attraction will be the resultant of the corresponding elementary attractions, and therefore the actual motion will be the resultant of all the motions arising from the elementary disturbances. We have therefore only to investigate the elementary disturbances one by one, and having established the theory of these, we calculate the actual motion by combining the series of motions so obtained.

Assuming the motion of the satellites in one of the elementary disturbances to be that of oscillation about a mean position, and the whole motion to be that of a uniformly revolving series of undulations, we find our supposition to be correct, provided a certain biquadratic equation is satisfied by the quantity denoting the rate of oscillation. § 6.

When the four roots of this equation are all real, the motion of each satellite is compounded of four different oscillations of different amplitudes and periods, and the motion of the whole ring consists of four series of undulations, travelling round the ring with different velocities. When any of these roots are impossible, the motion is no longer oscillatory, but tends to the rapid destruction of the ring.

To determine whether the motion of the ring is permanent, we must assure ourselves that the four roots of this equation are real, whatever be the number of undulations in the ring; for if any one of the possible elementary disturbances should lead to destructive oscillations, that disturbance might sooner or later commence, and the ring would be destroyed.

Now the number of undulations in the ring may be any whole number from one up to half the number of satellites. The forces from which danger is to be apprehended are greatest when the number of undulations is greatest, and by taking that number equal to half the number of satellites, we find the condition of stability to be

$$S > .4352\,\mu^3 R,$$

where S is the mass of the central body, R that of the ring, and μ the number of satellites of which it is composed. § 8. If the number of satellites be too great, destructive oscillations will commence, and finally some of the satellites will come into collision with each other and unite, so that the number of independent satellites will be reduced to that which the central body can retain and keep in discipline. When this has taken place, the satellites will not only be kept at the proper distance from the primary, but will be prevented by its preponderating mass from interfering with each other.

We next considered more carefully the case in which the mass of the ring is very small, so that the forces arising from the attraction of the ring are small compared with that due to the central body. In this case the values of the roots of the biquadratic are all real, and easily estimated. § 9.

If we consider the motion of any satellite about its mean position, as referred to axes fixed in the plane of the ring, we shall find that it describes an ellipse in the direction opposite to that of the revolution of the ring, the periodic time being to that of the ring as ω to n, and the tangential amplitude of oscillation being to the radial as 2ω to n. § 10.

The absolute motion of each satellite in space is nearly elliptic for the large values of n, the axis of the ellipse always advancing slowly in the direction of rotation. The path of a satellite corresponding to one of the small values of n is nearly circular, but the radius slowly increases and diminishes during a period of many revolutions. § 11.

The form of the ring at any instant is that of a re-entering curve, having m alternations of distance from the centre, symmetrically arranged, and m points of condensation, or crowding of the satellites, which coincide with the points of greatest distance when n is positive, and with the points nearest the centre when n is negative. § 12.

This system of undulations travels with an angular velocity $-\dfrac{n}{m}$ relative to the ring, and $\omega - \dfrac{n}{m}$ in space, so that during each oscillation of a satellite a complete wave passes over it. § 14.

To exhibit the movements of the satellites, I have made an arrangement by which 36 little ivory balls are made to go through the motions belonging to the first or fourth series of waves. (Figs. 7, 8.)

The instrument stands on a pillar A, in the upper part of which turns the cranked axle CC. On the parallel parts of this axle are placed two wheels, RR and TT, each of which has 36 holes at equal distances in a circle near its circumference. The two circles are connected by 36 small cranks of the form KK, the extremities of which turn in the corresponding holes of the two wheels. That axle of the crank K which passes through the hole in the wheel S is bored, so as to hold the end of the bent wire which carries the satellite S. This wire may be turned in the hole so as to place the bent part carrying the satellite at any angle with the crank. A pin P, which passes through the top of the pillar, serves to prevent the cranked axle from turning; and a pin Q, passing through the pillar horizontally, may be made to fix the wheel R, by inserting it in a hole in one of the spokes of that wheel. There is also a handle H, which is in one piece with the wheel T, and serves to turn the axle.

Now suppose the pin P taken out, so as to allow the cranked axle to turn, and the pin Q inserted in its hole, so as to prevent the wheel R from revolving; then if the crank C be turned by means of the handle H, the wheel T will have its centre carried round in a vertical circle, but will remain parallel to itself during the whole motion, so that every point in its plane will describe an equal circle, and all the cranks K will be made to revolve exactly as the large crank C does. Each satellite will therefore revolve in a small circular orbit, in the same time with the handle H, but the position of each satellite in that orbit may be arranged as we please, according as we turn the wire which supports it in the end of the crank.

In fig. 8, which gives a front view of the instrument, the satellites are so placed that each is turned 60^0 further round in its socket than the one behind it. As there are 36 satellites, this process will bring us back to our starting-point after six revolutions of the direction of the arm of the satellite; and therefore as we have gone round the ring once in the same direction, the arm of the satellite will have overtaken the radius of the ring five times.

Hence there will be five places where the satellites are beyond their mean distance from the centre of the ring, and five where they are within it, so that we have here a series of five undulations round the circumference of the ring. In this case the satellites are crowded together when nearest to the centre, so that the case is that of the *first* series of waves, when $m = 5$.

Now suppose the cranked axle C to be turned, and all the small cranks K to turn with it, as before explained, every satellite will then be carried round on its own arm in the same direction; but, since the direction of the arms of different satellites is different, their phases of revolution will preserve the same difference, and the system of satellites will still be arranged in five undulations, only the undulations will be propagated round the ring in the direction opposite to that of the revolution of the satellites.

To understand the motion better, let us conceive the centres of the orbits of the satellites to be arranged in a straight line instead of a circle, as in fig. 10. Each satellite is here represented in a different phase of its orbit, so that as we pass from one to another from left to right, we find the position of the satellite in its orbit altering in the direction opposite to that of the hands of a watch. The satellites all lie in a trochoidal curve, indicated by the line through them in the figure. Now conceive every satellite to move in its orbit through a certain angle in the direction of the arrows. The satellites will then lie in the dotted line, the form of which is the same as that of the former curve, only shifted in the direction of the large arrow. It appears, therefore, that as the satellites revolve, the undulation travels, so that any part of it reaches successively each satellite as it comes into the same phase of rotation. It therefore travels from those satellites which are most advanced in phase to those which are less so, and passes over a complete wave-length in the time of one revolution of a satellite.

Now if the satellites be arranged as in fig. 8, where each is more advanced in phase as we go round the ring in the direction of rotation, the wave will travel in the direction opposite to that of rotation, but if they are arranged as in fig. 12, where each satellite is less advanced in phase as we go round the ring, the wave will travel in the direction of rotation. Fig. 8 represents the *first* series of waves where $m = 5$, and fig. 12 represents the *fourth* series where $m = 7$. By arranging the satellites in their sockets before starting, we might make m equal to any whole number, from 1 to 18. If we chose any number above 18 the result would be the same as if we had taken a number as much below 18 and changed the arrangement from the first wave to the fourth.

In this way we can exhibit the motions of the satellites in the first and fourth waves. In reality they ought to move in ellipses, the major axes being twice the minor, whereas in the machine they move in circles: but the character of the motion is the same, though the form of the orbit is different.

We may now show these motions of the satellites among each other, combined with the motion of rotation of the whole ring. For this purpose we put in the pin P, so as to prevent the crank axle from turning, and take out the pin Q so as to allow the wheel R to turn. If we then turn the wheel T, all the small cranks will remain parallel to the fixed crank, and the wheel R will revolve at the same rate as T. The arm of each satellite will continue parallel to itself during the motion, so that the satellite will describe a circle whose centre is at a distance from the centre of R, equal to the arm of the satellite, and measured in the same direction. In our theory of real satellites, each moves in an ellipse, having the central body in its focus, but this motion in an eccentric circle is sufficiently near for illustration. The motion of the waves relative to the ring is the same as before. The waves of the first kind travel faster than the ring itself, and overtake the satellites, those of the fourth kind travel slower, and are overtaken by them.

In fig. 11 we have an exaggerated representation of a ring of twelve satellites affected by a wave of the fourth kind where $m = 2$. The satellites here lie in an ellipse at any given instant, and as each moves round in its circle about its mean position, the ellipse also moves round in the same direction with half their angular velocity. In the figure the dotted line represents the position of the ellipse when each satellite has moved forward into the position represented by a dot.

Fig. 13 represents a wave of the first kind where $m = 2$. The satellites at any instant lie in an epitrochoid, which, as the satellites revolve about their mean positions, revolves in the opposite direction with half their angular velocity, so that when the satellites come into the positions represented by the dots, the curve in which they lie turns round in the opposite direction and forms the dotted curve.

In fig. 9 we have the same case as in fig. 13, only that the absolute orbits of the satellites in space are given, instead of their orbits about their mean positions in the ring. Here each moves about the central body in an eccentric circle, which in strictness ought to be an ellipse not differing much from the circle.

As the satellites move in their orbits in the direction of the arrows, the curve which they form revolves in the same direction with a velocity $1\frac{1}{2}$ times that of the ring.

By considering these figures, and still more by watching the actual motion of the ivory balls in the model, we may form a distinct notion of the motions of the particles of a discontinuous ring, although the motions of the model are circular and not elliptic. The model, represented on a scale of one-third in figs. 7 and 8, was made in brass by Messrs. Smith and Ramage of Aberdeen.

We are now able to understand the mechanical principle, on account of which a massive central body is enabled to govern a numerous assemblage of satellites, and to space them out into a regular ring; while a smaller central body would allow disturbances to arise among the individual satellites, and collisions to take place.

When we calculated the attractions among the satellites composing the ring, we found that if any satellite be displaced tangentially, the resultant attraction will draw it away from its mean position, for the attraction of the satellites it approaches will increase, while that of those it recedes from will diminish, so that its equilibrium when in the mean position is unstable with respect to tangential displacements; and therefore, since every satellite of the ring is statically unstable between its neighbours, the slightest disturbance would tend to produce collisions among the satellites, and to break up the ring into groups of conglomerated satellites.

But if we consider the dynamics of the problem, we shall find that this effect need not necessarily take place, and that this very force which tends towards destruction may become the condition of the preservation of the ring. Suppose the whole ring to be revolving round a central body, and that one satellite gets in advance of its mean position. It will then be attracted forwards, its path will become less concave towards the attracting body, so that its distance from that body will increase. At this increased distance its angular velocity will be less, so that instead of overtaking those in front, it may by this means be made to fall back to its original position. Whether it does so or not must depend on the actual values of the attractive forces and on the angular velocity of the ring. When the angular velocity is great and the attractive forces small, the compensating process will go on vigorously, and the ring will be preserved. When the angular velocity is small and the attractive forces of the ring great, the dynamical effect will not compensate for the disturbing action of the forces and the ring will be destroyed.

If the satellite, instead of being displaced forwards, had been originally behind its mean position in the ring, the forces would have pulled it backwards, its path would have become more concave towards the centre, its distance from the centre would diminish, its angular velocity would increase, and it would gain upon the rest of the ring till it got in front of its mean position. This effect is of course dependent on the very same conditions as in the former case, and the actual effect on a disturbed satellite would be to make it describe an orbit about its mean position in the ring, so that if in advance of its mean position, it first recedes from the centre, then falls behind its mean position in the ring, then approaches the centre within the mean distance, then advances beyond its mean position, and, lastly, recedes from the centre till it reaches its starting-point, after which the process is repeated indefinitely, the orbit being always described in the direction opposite to that of the revolution of the ring.

We now understand what would happen to a disturbed satellite, if all the others were preserved from disturbance. But, since all the satellites are equally free, the motion of one will produce changes in the forces acting on the rest, and this will set them in motion, and this motion will be propagated from one satellite to another round the ring. Now propagated disturbances constitute waves, and all waves, however complicated, may be reduced to combinations of simple and regular waves; and therefore all the disturbances of the ring may be considered as the resultant of many series of waves, of different lengths, and travelling with different velocities. The investigation of the relation between the length and velocity of these waves forms the essential part of the problem, after which we have only to split up the original disturbance into its simple elements, to calculate the effect of each of these separately, and then to combine the results. The solution thus obtained will be perfectly general, and quite independent of the particular form of the ring, whether regular or irregular at starting. § 14.

We next investigated the effect upon the ring of an external disturbing force. Having split up the disturbing force into components of the same type with the waves of the ring (an operation which is always possible), we found that each term of the disturbing force generates a "forced wave" travelling with its own angular velocity. The magnitude of the forced wave depends not only on that of the disturbing force, but on the angular velocity with which the disturbance travels round the ring, being greater in proportion as this velocity more nearly coincides with that of one of the "free waves" of the ring. We also found that the displacement of the satellites was sometimes in the direction of the disturbing force, and sometimes in the opposite direction, according to the relative position of the forced wave among the four natural ones, producing in the one case positive, and in the other negative forced waves. In treating the problem generally, we must determine the forced waves belonging to every term of the disturbing force, and combine these with such a system of free waves as shall reproduce the initial state of the ring. The subsequent motion of the ring is that which would result from the free waves and forced waves together. The most important class of forced waves are those which are produced by waves in neighbouring rings. § 15.

We concluded the theory of a ring of satellites by tracing the process by which the ring would be destroyed if the conditions of stability were not fulfilled. We found two cases of instability, depending on the nature of the tangential force due to tangential displacement. If this force be in the direction opposite to the displacement, that is, if the parts of the ring are *statically stable*, the ring will be destroyed, the irregularities becoming larger and larger without being propagated round the ring. When the tangential force is in the direction of the tangential displacement, if it is below a certain value, the disturbances will be propagated round the ring without becoming larger, and we have the case of stability treated of at large. If the force exceed this value, the disturbances will still travel round the ring, but they will increase in amplitude continually till the ring falls into confusion. § 18.

We then proceeded to extend our method to the case of rings of different constitutions. The first case was that of a ring of satellites of unequal size. If the central body be of sufficient mass, such a ring will be spaced out, so that the larger satellites will be at wider intervals than the smaller ones, and the waves of disturbance will be propagated as before, except that there may be reflected waves when a wave reaches a part of the ring where there is a change in the average size of the satellites. § 19.

The next case was that of an annular cloud of meteoric stones, revolving uniformly about the planet. The *average density* of the space through which these small bodies are scattered will vary with every irregularity of the motion, and this variation of density will produce variations in the forces acting upon the other parts of the cloud, and so disturbances will be propagated in this ring, as in a ring of a finite number of satellites. The condition that such a ring should be free from destructive oscillations is, that the density of the planet should be more than three hundred times that of the ring. This would make the ring much rarer than common air, as regards its *average density*, though the density of the particles of which it is composed may be great. Comparing this result with Laplace's minimum density of a ring revolving as a whole, we find that such a ring cannot revolve as a whole, but that the inner parts must have a greater angular velocity than the outer parts. § 20.

We next took up the case of a flattened ring, composed of incompressible fluid, and moving with uniform angular velocity. The internal forces here arise partly from attraction and partly from fluid pressure. We began by taking the case of an infinite stratum of fluid affected by regular waves, and found the accurate values of the forces in this case. For long waves the resultant force is in the same direction as the displacement, reaching a maximum for waves whose length is about ten times the thickness of the stratum. For waves about five times as long as the stratum is thick there is no resultant force, and for shorter waves the force is in the opposite direction to the displacement. § 23.

Applying these results to the case of the ring, we find that it will be destroyed by the long waves unless the fluid is less than $\frac{1}{42}$ of the density of the planet, and that in all cases the short waves will break up the ring into small satellites.

Passing to the case of *narrow* rings, we should find a somewhat larger maximum density, but we should still find that very short waves produce forces in the direction opposite to the displacement, and that therefore, as already explained (page 333), these short undulations would increase in magnitude without being propagated along the ring, till they had broken up the fluid filament into drops. These drops may or may not fulfil the condition formerly given for the stability of a ring of equal satellites. If they fulfil it, they will move as a permanent ring. If they do not, short waves will arise and be propagated among the satellites, with ever increasing magnitude, till a sufficient number of drops have been brought into collision, so as to unite and form a smaller number of larger drops, which may be capable of revolving as a permanent ring.

We have already investigated the disturbances produced by an external force independent of the ring; but the special case of the mutual perturbations of two concentric rings is considerably more complex, because the existence of a double system of waves changes the character of both, and the waves produced react on those that produced them.

We determined the attraction of a ring upon a particle of a concentric ring, first, when both rings are in their undisturbed state; secondly, when the particle is disturbed; and, thirdly, when the attracting ring is disturbed by a series of waves. § 26.

We then formed the equations of motion of one of the rings, taking in the disturbing forces arising from the existence of a wave in the other ring, and found the small variation of the velocity of a wave in the first ring as dependent on the magnitude of the wave in the second ring, which travels with it. § 27.

The forced wave in the second ring must have the same absolute angular velocity as the free wave of the first which produces it, but this velocity of the free wave is slightly altered by the reaction of the forced wave upon it. We find that if a free wave of the first ring has an absolute angular velocity not very different from that of a free wave of the second ring, then if both free waves be of even orders (that is, of the second or fourth varieties of waves), or both of odd orders (that is, of the first or third), then the swifter of the two free waves has its velocity increased by the forced wave which it produces, and the slower free wave is rendered still slower by its forced wave; and even when the two free waves have the same angular velocity, their mutual action will make them both split into two, one wave in each ring travelling faster, and the other wave in each ring travelling slower, than the rate with which they would move if they had not acted on each other.

But if one of the free waves be of an even order and the other of an odd order, the swifter free wave will travel slower, and the slower free wave will travel swifter, on account of the reaction of their respective forced waves. If the two free waves have naturally a certain small difference of velocities, they will be made to travel together, but if the difference is less than this, they will again split into two pairs of waves, one pair continually increasing in magnitude without limit, and the other continually diminishing, so that one of the waves in each ring will increase in violence till it has thrown the ring into a state of confusion.

There are four cases in which this may happen. The first wave of the outer ring may conspire with the second or the fourth of the inner ring, the second of the outer with the third of the inner, or the third of the outer with the fourth of the inner. That two rings may revolve permanently, their distances must be arranged so that none of these conspiracies may arise between odd and even waves, whatever be the value of m. The number of conditions to be fulfilled is therefore very great, especially when the rings are near together and have nearly the same angular velocity, because then there are a greater number of dangerous values of m to be provided for.

In the case of a large number of concentric rings, the stability of each pair must be investigated separately, and if in the case of any two, whether consecutive rings or not, there are a pair of conspiring waves, those two rings will be agitated more and more, till waves of that kind are rendered impossible by the breaking up of those rings into some different arrangement. The presence of the other rings cannot prevent the mutual destruction of any pair which bear such relations to each other.

It appears, therefore, that in a system of many concentric rings there will be continually new cases of mutual interference between different pairs of rings. The forces which excite these disturbances being very small, they will be slow of growth, and it is possible that by the irregularities of each of the rings the waves may be so broken and confused (see § 19), as to be incapable of mounting up to the height at which they would begin to destroy the arrangement of the ring. In this way it may be conceived to be possible that the gradual disarrangement of the system may be retarded or indefinitely postponed.

But supposing that these waves mount up so as to produce collisions among the particles, then we may deduce the result upon the system from general dynamical principles. There will be a tendency among the exterior rings to remove further from the planet, and among the interior rings to approach the planet, and this either by the extreme interior and exterior rings diverging from each other, or by intermediate parts of the system moving away from the mean ring. If the interior rings are observed to approach the planet, while it is known that none of the other rings have expanded, then the cause of the change cannot be the mutual action of the parts of the system, but the resistance of some medium in which the rings revolve. § 31.

There is another cause which would gradually act upon a broad fluid ring of which the parts revolve each with the angular velocity due to its distance from the planet, namely, the internal friction produced by the slipping of the concentric rings with different angular velocities. It appears, however (§ 33), that the effect of fluid friction would be insensible if the motion were regular.

Let us now gather together the conclusions we have been able to draw from the mathematical theory of various kinds of conceivable rings.

We found that the stability of the motion of a solid ring depended on so delicate an adjustment, and at the same time so unsymmetrical a distribution of mass, that even if the exact condition were fulfilled, it could scarcely last long, and if it did, the immense preponderance of one side of the ring would be easily observed, contrary to experience. These considerations, with others derived from the mechanical structure of so vast a body, compel us to abandon any theory of solid rings.

We next examined the motion of a ring of equal satellites, and found that if the mass of the planet is sufficient, any disturbances produced in the arrangement of the ring will be propagated round it in the form of waves, and will not introduce dangerous confusion. If the satellites are unequal, the propagation of the waves will no longer be regular, but disturbances of the ring will in this, as in the former case, produce only waves, and not growing confusion. Supposing the ring to consist, not of a single row of large satellites, but of a cloud of evenly distributed unconnected particles, we found that such a cloud must have a very small density in order to be permanent, and that this is inconsistent with its outer and inner parts moving with the same angular velocity. Supposing the ring to be fluid and continuous, we found that it will be necessarily broken up into small portions.

We conclude, therefore, that the rings must consist of disconnected particles; these may be either solid or liquid, but they must be independent. The entire system of rings must therefore consist either of a series of many concentric rings, each moving with its own velocity, and having its own systems of waves, or else of a confused multitude of revolving particles, not arranged in rings, and continually coming into collision with each other.

Taking the first case, we found that in an indefinite number of possible cases the mutual perturbations of two rings, stable in themselves, might mount up in time to a destructive magnitude, and that such cases must continually occur in an extensive system like that of Saturn, the only retarding cause being the possible irregularity of the rings.

The result of long-continued disturbance was found to be the spreading out of the rings in breadth, the outer rings pressing outwards, while the inner rings press inwards.

The final result, therefore, of the mechanical theory is, that the only system of rings which can exist is one composed of an indefinite number of unconnected particles, revolving round the planet with different velocities according to their respective distances. These particles may be arranged in series of narrow rings, or they may move through each other irregularly. In the first case the destruction of the system will be very slow, in the second case it will be more rapid, but there may be a tendency towards an arrangement in narrow rings, which may retard the process.

We are not able to ascertain by observation the constitution of the two outer divisions of the system of rings, but the inner ring is certainly transparent, for the limb of Saturn has been observed through it. It is also certain, that though the space occupied by the ring is transparent, it is not through the material parts of it that Saturn was seen, for his limb was observed without distortion; which shows that there was no refraction, and therefore that the rays did not pass through a medium at all, but between the solid or liquid particles of which the ring is composed. Here then we have an optical argument in favour of the theory of independent particles as the material of the rings. The two outer rings may be of the same nature, but not so exceedingly rare that a ray of light can pass through their whole thickness without encountering one of the particles.

Finally, the two outer rings have been observed for 200 years, and it appears, from the careful analysis of all the observations by M. Struvé, that the second ring is broader than when first observed, and that its inner edge is nearer the planet than formerly. The inner ring also is suspected to be approaching the planet ever since its discovery in 1850. These appearances seem to indicate the same slow progress of the rings towards separation which we found to be the result of theory, and the remark, that the inner edge of the inner ring is most distinct, seems to indicate that the approach towards the planet is less rapid near the edge, as we had reason to conjecture. As to the apparent unchangeableness of the exterior diameter of the outer ring, we must remember that the outer rings are certainly far more dense than the inner one, and that a small change in the outer rings must balance a great change in the inner one. It is possible, however, that some of the observed changes may be due to the existence of a resisting medium. If the changes already suspected should be confirmed by repeated observations with the same instruments, it will be worth while to investigate more carefully whether Saturn's Rings are permanent or transitionary elements of the Solar System, and whether in that part of the heavens we see celestial immutability, or terrestrial corruption and generation, and the old order giving place to new before our own eyes.

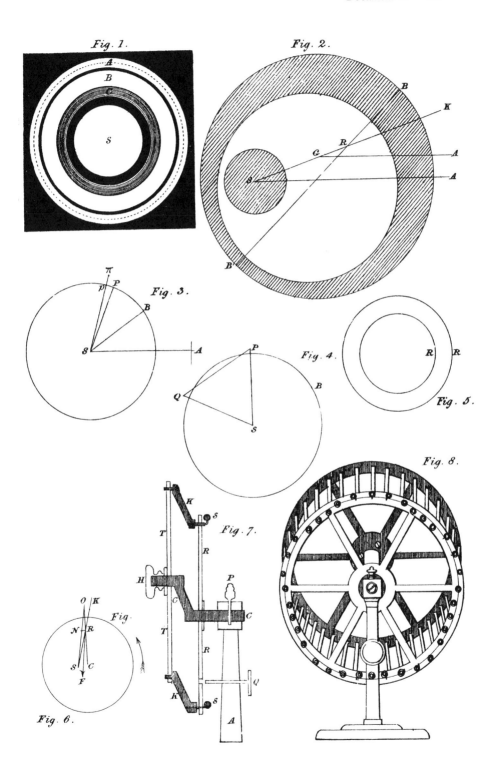

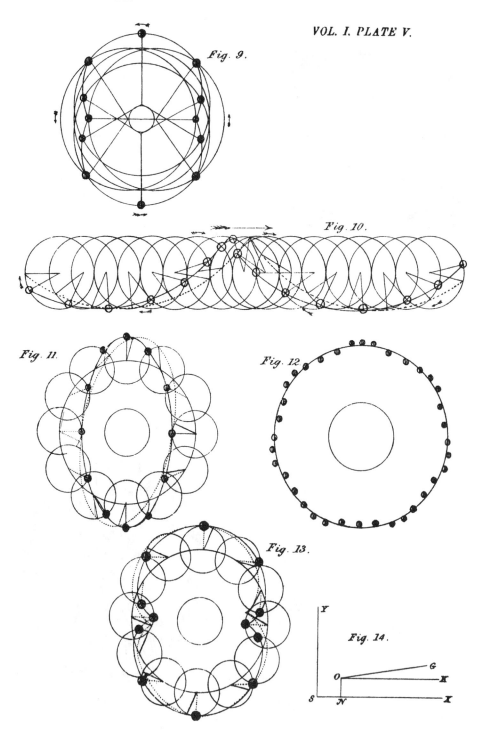

VOL. I. PLATE V.

Fig. 9.

Fig. 10.

Fig. 11.

Fig. 12.

Fig. 13.

Fig. 14.

18. George Biddell Airy's Review of Maxwell's Essay on Saturn's Rings[a]

Monthly Notices of the Royal Astronomical Society 19(June 10, 1859), 297–304.

The remarkable essay of which we have given the title was published in the beginning of the present year. The subject of it is so interesting, the difficulty of treating it in its utmost generality so considerable, and the results at which the author arrives so curious, that we think a brief abstract of it will be acceptable to the readers of the *Monthly Notices*. We shall commence with a very imperfect reference to preceding investigations on the same subject.

The first to which we shall allude is Laplace's, in the *Mécanique Céleste*, livre III chapitre vi.[b] Laplace considers a ring of *Saturn* as a solid, the form of which is investigated as if it were fluid (a mode of treatment whose result, in respect of the form of equilibrium, is evidently good for a solid), and finds, that if the breadth and thickness of the ring are very small in comparision with its distance from *Saturn*, its section may be an ellipse; and it appears that the formula for the proportion of the axes of the ellipse admits of its being considerably flattened. But Laplace rather inclines to the supposition that there are several rings, each existing by its own proper theory. Then remarking on the appearances noticed by some observers which seem to indicate irregularities in the rings, he adds, "J'ajoute que ces inégalités sont nécessaires pour maintenir l'anneau en équilibre autour de *Saturne*,"[c] and gives an investigation which shows that, if the ring were rigid and uniform, the slightest disturbance would cause it to fall to the planet. Then he concludes thus, "Les divers anneaux qui entourent le globe de *Saturne* sont, par conséquent, des solides irreguliers d'une largeur inégale dans les différents points de leurs circonférences,"[d] &c. And in this state he leaves the theory.

Probably no competent mathematician has ever read this paper without remarking the want of ground for the last conclusion. Professor Maxwell has well given the following as the conclusion that ought to have been drawn: "If the rings were solid and uniform, their motion would be unstable, and they would be destroyed. But they are not destroyed, and their motion is stable; therefore they are either not uniform or not solid."

Omitting notices of some remarks by Plana[e] and others which imply no important departure from Laplace's theory, we come to a paper by Mr. G. P. Bond, in the *Astronomical Journal*, Nos. 25 and 26.[f] In No. 25 Mr. Bond gives arguments from observation tending to show that there are several rings. He then gives investigations, in a great measure similar to Laplace's, but leading more distinctly to a determination of the limits of density and dimensions of any ring: he finds that the number of rings must be considerable, and even makes calculations for eleven rings. He then alludes (without further remark or calculation) to Laplace's ideas that a ring, if rigid, must be an irregular solid. And finally, he gives his opinion that a fluid symmetrical ring is not necessarily unstable.

In No. 27 Professor Benjamin Peirce has commented upon Mr. Bond's paper.[g] He "maintains, unconditionally, that there is no conceivable form of irregularity, and no combination of irregularities consistent with an actual ring, which would serve to retain it permanently about the primary if it were solid." He then refers to the theory of a fluid ring, and asserts that "there is nothing in the direct action of *Saturn* to prevent his ring from moving forward in its plane, in any direction and to any distance, until at last the rings would be brought into collision with the surface of the planet, and so be destroyed." Subsequently "the power which sustains the centre of gravity of *Saturn's* ring is not then to be sought in the planet itself, but in his satellites," and afterwards, "It follows, then, that no planet can have a ring, unless it is surrounded by a sufficient number of properly arranged satellites." This paper, it must be remarked, contains no symbolical investigations. Perhaps many mathematicians will think that a very complete mathematical treatment will be required to establish the two laws, that the attraction of *Saturn* will not support a fluid ring, and that the attraction of satellites will tend to support it.

In the same journal, No. 86, is a paper of mathematical character on the same subject by Professor Pierce.[h] But after some investigation, skilfully based on the method of potentials, as to the rotation of a solid ring, he refers merely to general reasoning in order to establish that "the solid ring must be excluded from any physical theory which rests upon a firm basis." He adds, however, "the conditions of permanence would lose the unstable element, if the two successive normals to the level surface of the rings [*quaere*, to the surface of equal potentials] at the centre of *Saturn* intersected each other between the centre of *Saturn* and the centre of gravity of the ring." It seems possible that this may refer to the case of a very large protuberance at one point of the ring; but the interpretation of the expression as it stands is somewhat obscure. The paper then proceeds to treat of a fluid ring, and terminates with, "The attractions of the opposite elements of the ring upon *Saturn* balance each other, and the ring has no tendency to move *Saturn*, neither has *Saturn* any tendency to give a motion of translation to the ring, or to check a translation which the ring may have received from other causes.

Hence the ring will yield to the slightest foreign action, and may pass through successive normal forms of possible equilibrium without being restrained by the action of *Saturn*. The fluid ring cannot then be regarded as one of real permanence without the aid of foreign support, although the action of the primary is not positively destructive to this, as it is to the solid ring." The inference that "the fluid ring cannot be regarded as one of real permanence without the aid of foreign support," is, we presume, to be limited strictly by the preceding sentence. Thus it permits one elliptic form to be changed, by foreign action, to another elliptic form, each being permanent so long as it is not disturbed by further foreign action. The equations and the conclusions as to attraction are exactly the same as they would be for a ring of satellites not perturbing each other. In any case, we do not perceive the advantage of "foreign support." The author, however, has spoken so briefly on the subject that it is possible that we have mistaken his meaning. An expectation was held out that this paper would be continued, but, so far as we are aware, no continuation has yet appeared.[i]

Professor Maxwell commences his treatise with an abstract of Laplace's principal conclusions. He then states that he has confined himself to those parts of the subject which bear upon the question of the permanence of a given form of motion, and adds,

"There is a very general and very important problem in dynamics, the solution of which would contain all the results of this essay and a great deal more. It is this, 'Having found a particular solution of the equations of motion of any material system, to determine whether a slight disturbance of the motion indicated by the solution would cause a small periodic variation, or a total derangement of the motion.' The question may be made to depend upon the conditions of a maximum or a minimum of a function of many variables, but the theory of the tests for distinguishing maxima from minima by the calculus of variations becomes so intricate when applied to functions of several variables, that I think it doubtful whether the physical or the abstract problem will be first solved."

Part I. is headed "On the motion of a rigid body of any form about a sphere," attention being confined to the motion in the plane of reference. Equations being formed (founded on the method of potentials), they are then employed in the following propositions.

Prob. 1. "To find the conditions under which a uniform motion of the ring is possible."

Prob. 2. "To find the equations of the motion when slightly disturbed." Symbols for the disturbance of co-ordinates are attached to the symbols for co-ordinates found by the last problem, the new symbols being supposed liable to change with t, and a series of linear differential equations of the second order is formed.

Prob. 3. "To reduce the three simultaneous equations of motion to the form of a single equation."

Prob. 4. "To determine whether the motion of the ring is stable or unstable, by means of the relations of the co-efficients A, B, C, [introduced in prob. 3]."

Problems 3 and 4 give the key of the author's general method. In problem 3, the symbols of operation and of quantity are separated, n being then written for d/dt. (This amounts to the same as assuming that the variation of any element may be represented by $\varepsilon^{nt+\alpha}$.) Equations are thus formed which exhibit no differential co-efficients. Then it is remarked that supposing the value of n found, it will consist generally of the sum of a real and an imaginary term. If the imaginary term $= 0$, the variations of elements will increase, with the time, indefinitely and without change of sign. If both terms have values, the variations of elements will be periodical, with constantly increasing co-efficients. If the real term $= 0$, the variations of elements will be periodical, with constant co-efficients. The last-mentioned case is evidently the only one which is compatible with the idea of stability. The relations of the co-efficients A, B, C, are therefore to be ascertained, which will insure that the values of n be purely imaginary. We commend these propositions to the study of the reader, as an interesting example of a beautiful method, applied with great skill to the solution of the difficult problems which follow.

Prob. 5 determines certain quantities required in the application of the theory to a rigid ring of constant or variable section: and Prob. 6 then takes up the question, "To determine the conditions of stability of the motion in terms of the coefficients f, g, h, [found in the last problem], which indicates the distribution of mass in the ring." On applying this to specific suppositions, it is found (1) that a ring of uniform section is unstable; (2) that a ring, thicker on one side than the other, and varying in section according to the simple law of sines, is unstable; (3) that a uniform ring loaded with a heavy particle at a point of its circumference may be stable, provided f be less than .8279 and greater than .8159 ($2f$ being the coefficient of $\cos \theta$ in the expansion of the expression for the section of the ring at every point by Fourier's theorem). On calculating the case of $f = .82$, it appears that the mass of the heavy particle is to the mass of the ring as 82 to 18. Such a law of structure evidently has no counterpart in *Saturn's* rings. It appears also that, in the case supposed, every variable co-ordinate may then be subject to two periodic variations, the period of one being 1.69 revolution of the ring, and that of the other being 3.25 revolution of the ring.

It appears then that, though a rigid ring is not impossible, yet it is impossible that *Saturn's* rings can be rigid rings.

Part II. treats "On the Motion of a Ring, the parts of which are not rigidly connected."

The ring considered is small in section, and nearly circular and uniform, and revolving with nearly uniform velocity. The variations from circular form, uniform section, and uniform velocity, are expressed by an appropriate notation. The first step is "To express the position of an element of a variable ring at a given time in terms of the original position of the element in the ring." Fourier's theorem being applied to this the problem becomes one of treating a number of separate variations, each of which is expressed by a circular function. Then the investigation is proposed "To find the magnitude and direction of the attraction between two elements of a disturbed ring." Having effected this, the author remarks, "We have found the expressions for the forces which act upon each member of a system of equal satellites, which originally formed a uniform ring, but are now affected with displacements depending on circular functions. If these displacements can be propagated round the ring in the form of waves with the velocity m/n, the quantities under the brackets will depend on t, and the complete expressions will be

$$\rho = A \cos(ms + nt + \alpha)$$

$$\sigma = B \sin(ms + nt + \beta)$$

$$\zeta = C \cos(ms + nt + \gamma)$$

[ρ, σ, ζ, being perturbations of three elements of the particle's place]. Let us find in what cases expressions such as these will be true, and what will be the result when they are not true."

The conditions are shown to depend upon the possible or imaginary character of roots of an equation, nearly as in the first section; and the author soon arrives at this conclusion: "So that a ring of satellites can always be rendered stable by increasing the mass of the central body and the angular velocity of the ring." As an instance, supposing that there are 100 satellites, then the mass of the planet must not be less than 4352 times the entire mass of the ring.

Then the problem is considered "To determine the nature of the motion when the system of satellites is of small mass compared with the central body." After treating it generally, the author considers,—*First,* the motion of a single satellite with reference to its mean or undisturbed place; and shows that the motion is elliptic, the centre of the ellipse being at the mean place of the satellite, the major axis being in the tangential direction, the time of revolution being different from the time of revolution of the ring, and the direction of revolution being opposite to that of the ring. *Secondly*, the condition of the ring of satellites at a given instant: the author shows that the form of the ring at any instant is that of a string of beads, arranged as a circle affected by *m* regular waves of transversal displacement at equal intervals round the circle: besides these, there are waves of condensation and rarefaction, in which, according to the relation of coefficients, the points of greatest distance from the centre may be points of greatest or of least condensation. *Thirdly*, the condition of the ring of satellites through a duration of time: it is shown that the waves in question travel, with a motion which, in regard to the ring itself, is opposite to the ring's revolution.

There is then given a solution of the general problem, "From the state and motion of the satellites at one time, to find the state and motion at any future time."

The next problem is, "To consider the effect of an external disturbing force." A result soon reached is, "If the absolute angular velocity of the disturbing body is exactly or nearly equal to the absolute angular velocity of any of the free waves of the ring, that wave will increase till the ring be destroyed."

For the understanding of further results, it is almost necessary to refer to the theory of tides and waves. It is known in that theory that the action of the moon produces on the ocean waves whose periods and lengths are both dependent on the moon alone (called *forced* waves), and that these are or may be accompanied with *free* waves, whose periods depend on the moon but whose lengths do not depend on the moon or whose lengths depend on the moon but whose periods do not depend on the moon: and also with other free waves, entirely independent of the moon. And thus in the ring of satellites there will be forced waves and free waves.

The problem is then considered, "On the motion of a ring of satellites, when the conditions of stability are not fulfilled;" and the directions in which they tend to depart from their mean positions are found.

Satellites of unequal mass, it is considered, may be so grouped in our contemplation of their movements as to be treated nearly in the same manner as equal satellites.

The case of a fluid ring is then considered, and the conclusion arrived at is this:—"It appears, therefore, that a ring composed of a continuous liquid mass cannot revolve about a central body without being broken up, but that the parts of such a broken ring may, under certain conditions, form a permanent ring of satellites."

The author then treats "On the mutual perturbations of two rings." In this investigation arises a most curious complication of eight waves in each ring, with corresponding periods for the waves of the two rings, but with different relations of the coefficients in the two rings.

The author alludes shortly to "the effect of long-continued disturbances on a system of rings," and also to "the loss of energy due to internal friction in a broad fluid ring, the parts of which revolve about the planet, each with the velocity of a satellite at the same distance," which he concludes to be insensible.

The most important paragraph in the author's recapitulation is the following:—

"The final result, therefore, of the mechanical theory is, that the only system of rings which can exist is one composed of an indefinite number of unconnected particles, revolving round the planet with different velocities according to their respective distance. These particles may be arranged in series of narrow rings, or they may move through each other irregularly. In the first case, the destruction of the system will be very slow; in the second case, it will be more rapid; but there may be a tendency towards an arrangement in narrow rings, which may retard the process.

"We are not able to ascertain by observation the constitution of the two outer divisions of the system of rings; but the inner ring is certainly transparent, for the limb of *Saturn* has been observed through it. It is also certain that, though the space occupied by the ring is transparent, it is not through the material parts of it that *Saturn* was seen, for his limb was observed without distortion; which shows that there was no refraction, and therefore that the rays did not pass through a medium at all, but between the solid or liquid particles of which the ring is composed. Here, then, we have a optical argument in favour of the theory of independent particles as the material of the rings. The two outer rings may be of the same nature, but not so exceedingly rare that a ray of light can pass through their whole thickness without encountering one of the particles."

We have omitted allusion to a geometrical representation, by mechanism, of a disturbed ring of satellites, and also of a theorem by Professor W. Thomson, which, being expressed in a different notation, cannot without difficulty be compared with the investigations of Professor Maxwell.

The abstract which we have given will, we think, fully justify the
opinion that the theory of *Saturn's* rings is now placed on a footing
totally different from any that it has occupied before, and that the essay
which we have abstracted is one of the most remarkable contributions
to mechanical astronomy that has appeared for many years.

a. George Biddell Airy (1801–1892) was appointed Astronomer Royal in 1835,
having previously held the Lucasian and Plumian professorships at Cambridge.
In the opinion of Olin J. Eggen, "Airy," *Dictionary of Scientific Biography*, vol.
1(1970), pp. 84–87, Airy deserves a place in that volume for his achievements in
organizing British astronomy rather than his own scientific achievements. He had
some influence on the choice of topics for the Adams Prize competition, as indicated
in the letter from Challis to Thomson, February 28, 1855, given in the introduction.

b. See the introduction, note 7.

c. "In addition these inequalities are necessary to maintain the ring in equilibrium
around Saturn."

d. "The several rings that surround the globe of *Saturn* are, consequently, made up
of irregular solids of unequal size at different points in their circumferences."

e. See the introduction, note 11.

f. See the introduction, note 16.

g. Benjamin Peirce, "On the Constitution of Saturn's Rings," *Astronomical Journal*
2(1852), 17–19; see also *American Journal of Science* 12(1851), 106–108.

h. P. 111 in B. Peirce, "On the Adams Prize-Problem for 1856," *Astronomical Journal*
4(1855), 110–112. The spelling of Peirce's name evidently defeated the English
printer on its second appearance in Airy's MS.

i. For Peirce's later work on Saturn's rings see the introduction, note 48.

19. Letter from Maxwell to George Bond, August 25, 1863[a]

*Memorials of William Cranch Bond, Director of the Harvard College
Observatory 1840–1859, and of his son, George Phillips Bond, Director of
the Harvard College Observatory 1859–1865*, Edward S. Holden (San
Francisco: Murdock, 1897), pp. 203–206. We are indebted to Dr. Owen
Gingerich for informing us of the existence of this letter.

Glenlair House, Dalbeattie, Scotland
25th, August, 1863

Dear Sir:—

I shall study what you say about *Saturn* in your letter when I see your
drawings and observations. I have no doubt that the time is coming
when we shall know more about the heavenly bodies than that they
attract each other from a distance. In *Saturn's* ring we certainly have a

very wonderful object to examine, and when we come to understand it we shall certainly know more mechanics then we do now.

Your observations of comet's tails go far to render them legitimate subjects of speculation, and I think that when we have mastered the theory of these tails we shall know more about what the heavens are made of. I think the heavenly spaces are by no means empty, since, as Thomson has shown, a cubic mile of sunlight, even at the earth's distance is worth, mechanically, 12,050 foot-pounds;[b] and a cubic foot of space near the sun can contain energy equal to .0038 foot-pounds *at least*. This is under ordinary circumstances, and gives an estimate of the amount of strain which the medium has been for ages subjected to, without in any way giving way. But we have no reason to believe that if the sun's heat were increased 1000-fold, the medium would be unable to transmit it, or would break down under the forces applied. We have therefore no knowledge of the ultimate strength of the heavenly medium; but it is well able to do all that is required of it, whether we give it nothing to do but to transmit light and heat, or whether we make it the machinery of magnetism and electricity also, and at last assign gravitation itself to its power.

If we could understand how the pressure of a dense body could produce a linear pressure radiating out in straight lines from the body, and keep up this kind of pressure continually, then gravitation would be explained on mechanical principles, and the attraction of two bodies would be the consequence of the repulsive action of the lines of pressure in the medium.

For instance, in the case of a body [P] at a distance from the sun [S] the equation to the lines of force would [diagram omitted]:[c]

$$P \cos \theta + r^2 \sin^2 \theta = C$$

Where r is the distance from P, and θ the angle which r makes with PS.

There are two sets of lines separated by the surface of revolution whose equation is got by making $C = Pr^2 = a^2/(1 + \cos \theta)$. This surface has the general shape of a paraboloid of revolution, but suggests the appearance of a comet's tail, being more like a catenary than a parabola near the head. Is there anything about a comet to render its lines of force visible, and not those of a planet which are stronger? I think that visible lines of gravitating force are extremely improbable, but I never saw anything so like them as some tails of comets. What Herschel says about the repulsive action of the sun leaves unexplained the fact that the motion of the nucleus is that of a body gravitating toward the sun with a force neither more nor less than that of ordinary matter. If there were at any time in the comet matter which was not gravitating, or not gravitating to the same extent as earthly matter, then the path of the comet would be less curved to the sun than if it were made of ordinary matter, and therefore calculations depending upon the common value of the sun's attractive power would not give the true path of comets.

I have nothing yet to send you, but we are making a report on electrical measurements for the British Association which I will send you when I get copies,[d] and if you will inform me of any electrical men in America, I will bring forward their claim to have copies of the *Standard Coil of Electrical Resistance.* We have hopes of producing coils next winter, the resistance of which is known to within a small fraction in electromagnetic units. Such coils may be employed in measuring electromotive forces, in determining the mechanical equivalent of heat, and in other researches. The present measures of resistance in absolute units vary by six or seven per cent, but I think we are already safe within one-half of one per cent, and I see how to make determinations quite as exact as we can determine the size of our coil in meters.

In the course of our work we have had to obtain a constant velocity of rotation. This was secured by means of a governor invented by Mr. Fleming Jenkin;[e] but we propose to make a new governor, combining the principles of Professor W. Thomson and Mr. Jenkin; we hope to get good results, comparable with clockwork. I have been studying the mathematical principles of governors, and I have been able to detect the sources of irregularities in the motion, and I hope to correct them. We mean to expose the new governor to severe tests by sudden variations of driving power, and if we find it answers I hope it will be taken into consideration in devising moving power for large equatorials. The dynamics of governors is exceedingly interesting, on account of the number of conditions which may be introduced by various arrangements of the machinery, and the different and sometimes opposite effects of these on the stability of the motion.[f]

I am exceedingly obliged to you for your kindness in sending the books. I hope to be able to say so again when I have read the part about *Saturn.* I think the visibility of the ring under oblique sunshine shows that its surface is very rough, the roughness not being like that of paper or sandstone, but like that of a wilderness of sharp rocks, so that we, being on the same side as the sun, see nearly every spot of sunshine while most of the shadows are hid by their respective objects. Arago's test of the solidity of a heavenly body by polarized light supposes the solid body to be as smooth as a rough bar of iron, if not actually polished, whereas the smoothest part of our earth is a paved street, and even the sea is generally too rough to polarize much light.

With much respect, yours truly,

J. Clerk Maxwell

a. George Phillips Bond (1825–1865): His whole life was centered on the Harvard Observatory, where he followed his father William Cranch Bond (1789–1859) as its director. There he made the observations that led him to the discovery of Hyperion, a satellite of Saturn, and to the crepe ring of the same planet.

b. William Thomson, "Note on the possible density of the Luminiferous Medium and on the Mechanical Value of a Cubic Mile of Sunlight," *Transactions of the Royal Society of Edinburgh* 21(1857), 57–61, reported in *Phil. Mag.* 9(1855), 36–40; the paper had been read in April 1854.

c. The notes in square brackets in the paragraph are in the original.

d. The report to which Maxwell refers is his "Report on the new unit of electrical resistance proposed and issued by the Committee on Electrical Standards appointed in 1861 by the British Association," *Proceedings of the Royal Society of London* 14(1865), 154–164; *Phil. Mag.* 29(1865), 477–486; *Ann. Physik* 126(1865), 369–387. Members of the committee included Maxwell, William Thomson, Balfour Stewart, and Fleming Jenkin.

e. Henry Charles Fleming Jenkin (1833–1885) was employed early in his life to design and manufacture submarine cable and cable–laying equipment. In addition to the above report Jenkin published supplementary ones with Maxwell. See Maxwell and Jenkin, "Description of further experimental measurement of electrical resistance made at Kings College," *British Association Report* (1864), pp. 350–351, and "On the elementary relations between electrical measurements," *Phil. Mag.* 29(1865), 436–460, 507–525. See also his biography by Robert Louis Stevenson.

f. Maxwell was drawn to the general investigation of the action of governors, on which he published, "On Governors," *Proceedings of the Royal Society of London* 16(1868), 270–283, reprinted in *Scientific Papers*, vol. 2, pp. 105–120. See also Otto Mayr, "Victorian Physicists and Speed Regulation: An Encounter between Science and Technology," *Notes and Records of the Royal Society of London* 26(1971), 205–228, and "Maxwell and the Origins of Cybernetics," *Isis* 62(1971), 425–444; A. T. Fuller, "The Early Development of Control Theory, II." *Transactions of the ASME/Journal of Dynamic Systems, Measurement and Control* 98(1976), 224–235.

20. [Theory of Saturn's Rings]

Cambridge University Library, Maxwell Manuscripts.

The existence of a thin flat and circular system of rings, surrounding, yet nowhere touching the planet Saturn has been known for 200 years but we have as yet little knowledge of their structure and little foundation for coming to a decision whether we may expect them to continue in their present ⟨position⟩ state for a long time, or whether their total or partial destruction is likely to be witnessed by living astronomers.

Considerable changes have certainly taken place in their appearance as seen by successive astronomers but we know that the earlier observers had more imperfect telescopes than those ⟨with⟩ which are now directed to Saturn so that the evidence requires very careful examination before we can conclude in favour of an actual change of form or a development of new features.

Mathematicians, however, while they accept the results of observation as to the changeableness or permanence of an astronomical phenomenon, are not at liberty to accept them as ultimate facts from which the future ⟨state⟩ phases of the motion may be deduced by the principle of continuity. They cannot regard Saturn's Ring as the result of some undiscovered law of generation and development prevailing among a class of planets. They must either explain on mechanical principles how they have continued to exist as long and whether they are ⟨likely⟩ subject to decay, or they must confess that a gap has been discovered in celestial mechanics which may be perceived by the telescope but cannot be stopped up by the calculus.

We ⟨are at liberty⟩ must suppose the Rings to consist of matter in one of the states known to us and to be acted on by gravitation. Laplace was the first to show that a solid uniform ring is necessarily unstable and must fall on the body of the planet.[a] I have shown that a single thin ring loaded at one point of its circumference with a mass rather more than $4\frac{1}{2}$ times its own weight *might* permanently revolve about a central body. ⟨Besides the extreme delicacy of the adjustment required⟩ The adjustment of weight however must be very accurate it must not be less than .8159 or greater then .8279 of the whole. The magnitude of such a weight would render it ⟨conspicuous⟩ impossible to escape observation and the delicacy of adjustment would soon be impaired by the immense forces called into play ⟨acting⟩ tending to break or bend the ring—forces under which the strongest materials known to us would behave like sand or wax.

The hypothesis of a ring forming a solid mass is therefore untenable. That of the coexistence of many such rings has still less mechanical possibility. We are therefore obliged to regard the rings as consisting of matter ⟨not⟩ the parts of which are not rigidly connected.

I have shown that it is possible for a ring consisting of a single row of unconnected particles to revolve permanently about the planet under certain conditions and that many such rings may revolve concentrically about the planet but that their mutual perturbations will gradually increase till some of them are thrown into a state of confusion.

The conclusion at which I arrived in my former paper was—"that the only system of rings which can exist is one composed of an indefinite number of unconnected particles revolving round the planet with different velocities according to their respective distances. These particles may be arranged in a series of narrow rings or they may move through each other irregularly. In the first case the destruction of the system will be very slow; in the second case it will be more rapid, but there may be a tendency towards an arrangement in narrow rings which may retard the process."[b]

I was then of the opinion that "When we come to deal with collisions among bodies of unknown numbers size and shape we can no longer trace the mathematical laws of their motions with any distinctness." (§(32))[c] I propose now to take up the question of this point and to endeavour to throw some light on the theory of a confused assemblage of jostling masses whirling round a large central body.

I shall not enter into the theory of the *formation* of a ring (see papers by Mr. Daniel Vaughan   ).[d] I shall suppose the rings already existing and find the conditions of their stability and the rate of their decay. In my former paper I restricted myself to cases in which no collisions take place and in which the mutual gravitation of the planet and the particles is the only force in action. It appears however that in this case that particles of each ring must be at a certain distance apart and each ring of particles must be at a considerable distance from the next so that even at the immense distances from which we view them we should see the discontinuity of their structure by their almost perfect transparency and the feebleness of their illumination. Whatever, therefore, may be the condition of the dark inner rings, the outer rings are too substantial to have a constitution of this kind. The particles composing them must be so near together that they influence each other far more by collisions and jostling than by the attraction of gravitation. The individual masses, even if large compared with meteoric stones or even mountains are probably so small compared with planets that the effect of the mutual gravitation of two of them on the velocity or path of either may be entirely neglected. We have therefore to consider the motions of an immense number of small bodies occupying a space in the form of a flat ring or rings of which the thickness is less than a thousandth of the diameter and whirling round a planet in the centre with velocities nearly corresponding to their respective distances.

Collisions will occur between these bodies and after collision each body will be projected with a velocity which will carry it into some other part of the cloud of particles, where ⟨the⟩ it will meet with other particles moving with a velocity different from its own. Another collision will thus occur and in this way the jostling of the particles once begun will be carried on throughout the system and kept up on account of the different velocities of the different parts of the system so as to produce a continual loss of energy and a decay in the motion of the ring.

The principles by which problems of this kind can be treated were first discussed by Prof Clausius in a paper "On the nature of the motions which we call heat,"[e] and were applied to several cases in gaseous physics by myself in a paper on the Motions and Collisions of Perfectly Elastic Spheres.[f] Professor Clausius[g] has since pointed out some mistakes in the latter parts of this paper in his paper on the conduction of Heat in Gases. I hope to be able to complete a correct investigation of diffusion and conduction of heat in gases and to establish the distribution of velocities among the particles in all cases, but at present I must confine my attention to the effects of the collisions of rough imperfectly elastic bodies, in which case the complete theory is much more difficult and we must be satisfied with certain approximations. This is less to be regretted as we do not know the coefficient of elasticity for the collisions of the pieces of Saturn's Rings and therefore we can ⟨get⟩ expect only approximate numerical results.

We compare the motions of any particle at any instant with that which it would have if it revolved uniformly about the central body at that distance. We find that it differs from it by a small quantity which may be resolved into three components, a radial component (x) and a tangential component (y) and a component normal to the orbit (z). These components constitute the *velocity of agitation* of the particle as distinguished from the velocity of the ring at that point which is that corresponding to a circular orbit. If we trace this velocity of agitation for a single particle we find that it ⟨describes⟩ moves among the different concentric rings as if it described an ellipse in the same time as the time of revolution but in the opposite direction the major axis being in the direction of the radius vector and equal to twice the minor. It also oscillates perpendicularly to the plane of the ring in a period rather less than that of revolution.

This is what would take place if the particle did not meet with any other particles in its course, but we know that other particles exist and can easily calculate the chance of its not being struck for a given time. In this way we can deduce an equation connecting the velocity of agitation of particles at the instant of their projection with that which they have when they suffer their next collision. On account of the difference of ⟨motion⟩ mean motion of the different concentric rings the motion of agitation among the particles when they meet is greater than that with which each was originally projected in the ring from which it came so that if the particles were perfectly elastic the motion of agitation would increase constantly till the rings were dispersed in a cloud. But if the particles are inelastic the velocities after collision are generally less than before it, and thus the increase in the motion of agitation due to the interpenetration of particles belonging to different rings, and the decay of the motion of agitation due to the imperfect elasticity of the whirling particles, may produce a kind of equilibrium or steadiness of motion in certain cases, while in others the motion of agitation will constantly increase or diminish, till some change is effected in the arrangement of the system.

In studying the nature of the motion of agitation we shall find that it does not consist of a system of disturbances in the motion distributed alike in all directions but that the directions and velocities are distributed differently in different directions. The measure of the agitation in a given direction which we shall adopt is found by multiplying half the mass of each particle by the square of the motion of agitation resolved in that direction. It appears that the energy of agitation is greatest in one direction, least in a direction at right angles to this and of intermediate value in a direction normal to the plane of these two directions, and that ⟨the⟩ its value in any other direction is found by the same rules as are used to determine "moments of inertia" and other mathematical quantities having an arrangement similar to that of the diameters of an ellipsoid. The sum of the energies in three rectangular directions is always equal to the total energy of agitation and the absolute energy is equal to the total energy of agitation together with the total energy due to the motion of the particles in mass.

We have next to examine the effect of collisions on the distribution of the agitation. We find that the total energy of agitation is diminished in the ratio of p to 1 where p has a value depending on the elasticity and on the nature of the bodies. The difference of the energy of agitation in any two of the principal axes is also diminished in the ratio of q to 1. The investigation of the value of p and q must be attended to separately.[h]

a. Laplace analyzed the stability of Saturn's rings in a paper published in 1789, and in more detail in *Traité de Mécanique Céleste* (1799), vol. II (see the introduction). Laplace concluded only that a uniform ring was unstable; therefore the ring must be an inhomogeneous solid of unequal widths around the circumference with the center of gravity not coincident with its geometric center.

b. Maxwell, "Saturn's Rings," *Scientific Papers*, vol. 1, p. 373.

c. Ibid. p. 354.

d. Daniel Vaughan (1818–1879) was born in Ireland and emigrated to the United States in 1840. He published papers on chemistry as well as astronomy. He retired from the chair in chemistry at Cincinnati College of Medicine and Surgery in 1872 and died in poverty seven years later (see the introduction and notes 51 and 52).

e. Asterisked in the original, but no footnote is entered at the bottom of the page. The paper referred to is Clausius, "On the kind of motion we Call Heat," *Ann Physik* 100(1857), 353–380, translated in *Phil. Mag.* 14(1857), 108–127.

f. Asterisked in the original, but no footnote is entered at the bottom of the page. The paper referred to is Maxwell, "Illustrations of the Dynamical Theory of Gases," *Phil. Mag.* 19(1860), 19–32, and 20(1860), 21–37, reprinted in *Scientific Papers*, vol. 1, pp. 377–409.

g. Asterisked in the original, but no footnote is entered at the bottom of the page. The paper referred to is Clausius, "On the Conduction of Heat by Gases," *Phil. Mag.* 24(1862), 417–435, 512–534.

h. Maxwell's ideas on using his methods in kinetic theory to the theory of Saturn's rings have been applied in A. F. Cooke and F. A. Franklin, "Rediscussion of Maxwell's Adams Prize Essay on the Stability of Saturn's Rings," *Astronomical Journal* 69(1964), 173–200.

21. "To find the relations between the velocities and rotations of two bodies of any form before and after impact"

Cambridge University Library, Maxwell Manuscripts.

I have considered the case of the collisions of perfectly elastic bodies of any form in a paper on the Dynamical Theory of Gases, (Philosophical Magazine July 1860) and have shown that the average vis viva of translation of every particle tends to become equal after many such collisions, and that the vis viva of rotation of each particle about each of its principal axes is equal and that the whole vis viva of rotation of each particle is equal to its vis viva of translation. The equality of the vis viva of particles of different sizes leads to ⟨the⟩ an explanation of Gay Lussac's law of atomic volumes of gases and the relation between the vis viva of translation and rotation leads to the result that Bernoulli's hypothesis in its simplest form will not explain the relation between the specific heat of air at constant pressure and at constant volume.

When the particles are not ⟨supposed⟩ perfectly elastic, the motion of agitation cannot be kept up without some external cause and the investigation of the question becomes much more complicated than in the case where the motion of agitation if ⟨confined⟩ the particles are confined within an elastic vessel is self sustaining and capable of attaining a constant state. The external cause ⟨of the⟩ which sustains the motion of agitation in the case of Saturn's rings is the different velocities of contiguous portions of the rings and the energy which is lost by collison is made up by a supply drawn partly from the energy of motion of the rings round Saturn and partly from the potential energy of their gravitation towards him. As this source of energy is gradually diminished the form of the rings is gradually altered.

As we cannot investigate the effects of the collisions of inelastic bodies as easily as when the elasticity is perfect and as our only object is to obtain approximate numerical values for p and q we shall simplify the calculations by the following assumptions.

1st That the principal moments of inertia of each body are equal and represented by $M_1 k_1^2$ and $M_2 k_2^2$.

2nd That the points of the two bodies at which the impact takes place are at distances a_1 and a_2 from their respective centres of gravity.

3rd That at a certain stage of the collision these two points are at rest relatively to each other on account of the action of the "impulse ⟨force⟩ of compression" which we may call R,

4th That the whole impulsive force acting between the two bodies consists of the impuls⟨ive force⟩ of compression R, and the impuls⟨ive force⟩ of restitution R' and that these two forces act in the same direction and are in the ratio of 1 to e where e is the coefficient of elasticity of impact for the two bodies.

It is this part of our assumption which is most precarious. It supposes that there is no slipping between the bodies and that the value of e is the same for normal and for oblique impact. Is is probable that the tangential impulse, when much smaller than the normal is independent of it but when the normal impulse is small the tangential impulse will have a maximum value equal to the normal impulse multiplied by a coefficient of impulsive friction.

In an investigation like the present in which we know so little about the bodies in motion it would not be advisable to begin by the introduction of complicated conditions arising from laws of friction and collision which at best are empirical.

We therefore begin with two rough spherical masses M_1 and M_2 whose radii are a_1 and a_2 and radii of gyration k_1 and k_2. Let the direction cosines of the line drawn from M_1 and M_2 at collision be l, m, n.

Let the velocities of the centres of gravity resolved along the axes of x, y, z be

u_1, v_1, w_1 and u_2, v_2, w_2	before impact
$\bar{u}_1, \bar{v}_1, \bar{w}_1$ and $\bar{u}_2, \bar{v}_2, \bar{w}_2$	at greatest compression
u_1', v_1', w_1' and u_2', v_2', w_2'	after restitution

Let the angular velocities of rotation about the axes x, y, z be denoted according to the same system of accentuation and suffixes by

$p, q, r.$

Let the components of the impulse of compression be

$X, Y, Z,$

and those of the impulse of restitution according to our hypothesis

$eX, eY, eZ,$

those of the whole impulse will be

$(1 + e)X, (1 + e)Y, (1 + e)Z.$

The equations of motion are now easily written down, but as their number is great we may confine ourselves to those relating to the axis of x and to the instant of greatest compression

$$\bar{u}_1 = u_1 + \frac{1}{M_1}X \qquad \bar{p}_1 = p_1 + \frac{a_1}{M_1 k_1{}^2}(mZ - nY)$$

$$\bar{u}_2 = u_2 + \frac{1}{M_2}X \qquad \bar{p}_2 = p_2 + \frac{a_2}{M_2 k_2{}^2}(mZ - nY)$$

To find the final velocities we must substitute an accent for the bar and multiply the impulse by $(1 + e)$.

The velocities of the striking points in the directions of x are

$$u_1 - (mr_1 - nq_1)a_1 \quad \text{and} \quad u_2 + (mr_2 - nq_2)a_2.$$

At the instant of greatest compression these points have no relative motion, or

$$\bar{u}_2 - \bar{u}_1 + m(a_1 r_1 + a_2 r_2) - n(a_1 \bar{q}_1 + a_2 \bar{q}_2) = 0.$$

Substituting the values of these velocities we find

$$\left\langle X\left(\frac{1}{M_1} + \frac{1}{M_2} + (m^2 + n^2)\frac{a_1{}^2}{M_1 k_1{}^2} + \frac{a_2{}^2}{M_2 k_2{}^2} - \right.\right\rangle$$

$$\left(\frac{1}{M_1} + \frac{1}{M_2}\right)X + \left(\frac{a_1{}^2}{M_1 k_1{}^2} + \frac{a_2{}^2}{M_2 k_2{}^2}\right)((m^2 + n^2)X - lmY - nlZ)$$

$$= u_2 - u_1 + m(a_1 r_1 + a_2 r_2) - n(a_1 q_1 + a_2 q_2)$$

with two other equations related to y and z as this is to x. Multiplying the first by

$$\left(\frac{1}{M_1} + \frac{1}{M_2} + l^2\left(\frac{a_1{}^2}{M_1 k_1{}^2} + \frac{a_2{}^2}{M_2 k_2{}^2}\right)\right)$$

the second by

$$lm\left(\frac{a_1{}^2}{M_1 k_1{}^2} + \frac{a_2{}^2}{M_2 k_2{}^2}\right)$$

and the third by

$$nl\left(\frac{a_1{}^2}{M_1 k_1{}^2} + \frac{a_2{}^2}{M_2 k_2{}^2}\right)$$

and adding we obtain the value of X.

$$X = \frac{1}{\dfrac{a_1{}^2 + k_1{}^2}{M_1 k_1{}^2} + \dfrac{a_2{}^2 + k_2{}^2}{M_2 k_2{}^2}}\left\{u_2 - u_1 + \frac{\dfrac{a_1{}^2}{M_1 k_1{}^2} + \dfrac{a_2{}^2}{M_2 k_2{}^2}}{\dfrac{1}{M_1} + \dfrac{1}{M_2}}l[l(u_2 - u_1)\right.$$

$$\left. + m(v_2 - v_1) + n(w_2 - w_1)] + m(a_1 r_1 + a_2 r_2) - n(a_1 q_1 + a_2 q_2)\right\}$$

In the same way we find

$$mZ - nY = \frac{1}{\dfrac{a_1{}^2 + k_1{}^2}{M_1 k_1{}^2} + \dfrac{a_2{}^2 + k_2{}^2}{M_2 k_2{}^2}}\{m(w_2 - w_1) - n(v_2 - v_1)$$

$$- (m^2 + n^2)(a_1 p_1 + a_2 p_2) + lm(a_1 q_1 + a_2 q_2)$$

$$+ ln(a_1 r_1 + a_2 r_2)\}$$

⟨To find the final velocities we must substitute these numbers in the⟩
If we write

$$\frac{a_1{}^2 + k_1{}^2}{M_1 k_1{}^2} + \frac{a_2{}^2 + k_2{}^2}{M_2 k_2{}^2} = \frac{2}{A}$$

and

$$\frac{\dfrac{a_1{}^2}{M_1 k_1{}^2} + \dfrac{a_2{}^2}{M_2 k_2{}^2}}{\dfrac{1}{M_1} + \dfrac{1}{M_2}} = B$$

in these expressions and substitute them in the equations for the final velocities,

$$u_1' = u_1 + \frac{1+e}{M_1}X \qquad p_1' = p_1 + \frac{(1+e)a_1}{M_1 k_1{}^2}(mZ - nY)$$

$$u_1' = u_1 + \frac{1+e}{M_1}\frac{A}{2}\{u_2 - u_1 + Bl(l(u_2 - u_1) + m(v_2 - v_1)$$

$$+ n(w_2 - w_1)) + m(a_1 r_1 + a_2 r_2) - n(a_1 q_1 + a_2 q_2)\}$$

$$p_1' = p_1 + \frac{(1+e)a_1}{M_1 k_1{}^2}\frac{A}{2}\{m(w_2 - w_1) - n(v_2 - v_1)$$

$$- (m^2 + n^2)(a_1 p_1 + a_2 p_2) + lm(a_1 q_1 + a_2 q_2) + ln(a_1 r_1 + a_2 r_2)\}$$

The other ten equations for the components of velocity and rotation may be easily written down from these.

We have next to determine the relations between the energy of agitation before and after impact. For this purpose we suppose the coordinate axes to be taken so as to coincide with the principal axes of agitation. We then square the equations for u_1' and p_1' remembering that ⟨the only terms which⟩ all terms containing products of different components of the velocities will disappear on summation, so that in obtaining mean values we retain only those terms containing squares of velocities. The following equations are therefore true of the mean squares of the quantities

$$u_1'^2 = u_1{}^2 - \frac{1+e}{M_1}A\{1 + Bl^2\}u_1{}^2 + \frac{(1+e)^2 A^2}{4M_1{}^2}\{(1 + 2Bl^2 + B^2 l^4)$$

$$\times (u_1{}^2 + u_2{}^2) + B^2 l^2 m^2(v_2{}^2 + v_1{}^2) + B^2 l^2 n^2(w_1{}^2 + w_2{}^2)$$

$$+ m^2(a_1{}^2 r_1{}^2 + a_2{}^2 r_2{}^2) + n^2(a_1{}^2 q_1{}^2 + a_2{}^2 q_2{}^2)\}$$

$$p_1'^2 = p_1{}^2 - \frac{(1+e)a_1{}^2}{M_1 k_1{}^2}A(m^2 + n^2)p_1{}^2 + \frac{(1+e)^2 A^2 a_1{}^2}{4M_1{}^2 k_1{}^4}\{m^2(w_1{}^2 + w_2{}^2)$$

$$+ n^2(v_1{}^2 + v_2{}^2) + (m^4 + 2m^2 n^2 + n^4)(a_1{}^2 p_1{}^2 + a_2{}^2 p_2{}^2)$$

$$+ l^2 m^2(a_1{}^2 q_1{}^2 + a_2{}^2 q_2{}^2) + l^2 n^2(a_1{}^2 r_1{}^2 + a_2{}^2 r_2{}^2)\}$$

Now l, m, n, the direction cosines of the line of centres at impact are independent of the velocities and by integrating over the surface of a sphere we find that the mean value of the square of a cosine such as l^2 is 1/3, that of the fourth power as l^4 is 1/5 and that of the square of a product as $l^2 m^2$ is 1/15. The equations may therefore be written

$$u_1'^2 = u_1^2 - \frac{(1+e)A}{M_1}(1 + \tfrac{1}{3}B)u_1^2 + \frac{(1+e)^2 A^2}{4M_1^2}\{(1 + \tfrac{2}{3}B + \tfrac{1}{5}B^2)$$

$$\times (u_1^2 + u_2^2) + \tfrac{1}{15}B^2(v_1^2 + v_2^2 + w_1^2 + w_2^2)$$

$$+ \tfrac{1}{3}(a_1^2 r_1^2 + a_2^2 r_2^2 + a_1^2 q_1^2 + a_2^2 q_2^2)\}$$

$$p_1'^2 = p_1^2 - \frac{2}{3}\frac{(1+e)Aa_1^2}{M_1 k_1^2}p_1^2 + \frac{(1+e)^2 A^2 a_1^2}{4M_1^2 k_1^2}\{\tfrac{1}{3}(v_1^2 + v_2^2 + w_1^2$$

$$+ w_2^2) + \tfrac{8}{15}(a_1^2 p_1^2 + a_2^2 p_2^2)$$

$$+ \tfrac{1}{15}(a_1^2 q_1^2 + a_2^2 q_2^2 + a_1^2 r_1^2 + a_2^2 r_2^2)\}$$

with ten other equations which may be written down from symmetry. When, as in the case of Saturn's rings, the motion of agitation is sustained by a cause which affects the motions of translation without directly altering the velocity of rotation ⟨we must have⟩ the ⟨motion⟩ energy of rotation must be sustained ⟨indirectly⟩ by the ⟨motion⟩ energy of translation and we must have $p_1'^2 = p_1^2$. The second equation then becomes

$$M_1 k_1^2 p_1^2 = \frac{(1+e)A}{8}\{v_1^2 + v_2^2 + w_1^2 + w_2^2 + \tfrac{8}{5}(a_1^2 p_1^2 + a_2^2 p_2^2)$$

$$+ \tfrac{1}{5}(a_1^2 q_1^2 + a_2^2 q_2^2 + a_1^2 r_1^2 + a_2^2 r_2^2)\}$$

Since ⟨this expression⟩ the right hand side of this expression would remain the same if the suffixes were exchanged, we must have

$$M_1 k_1^2 p_1^2 = M_2 k_2^2 p_2^2 = P\text{—suppose}$$

or the energy of rotation about the axis of x is the same for large and small particles. Let

$$\frac{a_1^2}{M_1 k_1^2} + \frac{a_2^2}{M_2 k_2^2} = 2C$$

then if we write $Mk^2 q^2 = Q$ and $Mk^2 r^2 = R$ we find

$$P = \frac{(1+e)A}{8}\{v_1^2 + v_2^2 + w_1^2 + w_2^2 + \tfrac{2}{5}C(8P + Q + R)\}$$

If we write down the two similar equations and add, we shall have

$$P + Q + R = \frac{(1+e)A}{4}\{u_1^2 + v_1^2 + w_1^2 + u_2^2 + v_2^2 + w_2^2$$

$$+ 2C(P + Q + R)\}$$

⟨Substituting the values of A and C we find and⟩ writing V_1^2 and V_2^2 for the squares of the resultant velocities

$$2(P + Q + R)(2 - (1 + e)AC) = (1 + e)A(V_1^2 + V_2^2)$$

$$\left\langle \left(\frac{2}{A} - 2eC \right)(P + Q + R)(1 + e)(u_1^2 + v_1^2 + w_1^2 + u_2^2 + v_2^2 + w_2^2) \right.$$

$$= \left\{ \frac{(1 - e)a_1^2 + k_1^2}{M_1 k_1^2} + \frac{(1 - e)a_2^2 + k_2^2}{M_2 k_2^2} \right\}(P + Q + R)$$

$$\left. = (1 + e)(V_1^2 + V_2^2) \right\rangle$$

When $e = 1$ this expression becomes

$$\langle (u_1^2 + v_1^2 + w_1^2 + u_2^2 + v_2^2 + w_2^2) \rangle = k_1^2(p_1^2 + q_1^2 + r_1^2)$$

$$+ k_2^2(p_2^2 + q_2^2 + r_2^2)$$

$$= V_1^2 + V_2^2$$

When $e = 0$ it becomes

$$\langle (u_1^2 + v_1^2 + w_1^2 + u_2^2 + v_2^2 + w_2^2) \rangle = \frac{2}{A}(P + Q + R) = V_1^2 + V_2^2$$

Subtracting the equation for Q from that for P we get

$$(P - Q)\left(\frac{2}{A} - \frac{7}{10}(1 + e)C \right) = \frac{1 + e}{4}(v_1^2 + v_2^2 - (u_1^2 + u_2^2))$$

Returning to the equation for $u_1'^2$ and adding the two similar equations

$$V_1'^2 = V_1^2 - \frac{(1 + e)A}{M_1}(1 + \tfrac{1}{3}B)V_1^2 + \frac{(1 + e)^2 A^2}{4M_1^2}(1 + \tfrac{2}{3}B + \tfrac{1}{3}B^2)$$

$$\times (V_1^2 + V_2^2) + \frac{(1 + e)^3 A^3 C}{6M_1^2(2 - (1 + e)AC)}(V_1^2 + V_2^2)$$

which gives the value of $V_1'^2$ in terms of V_1^2 and V_2^2. As we do not know the relation between V_1^2 and V_2^2 we shall begin with the case of equal masses and $V_1^2 = V_2^2$. Then we have

$$A = \frac{Mk^2}{a^2 + k^2} \qquad B = a^2/k^2 \qquad C = a^2/Mk^2$$

and if we make $V'^2 = p'V^2$ we find

$$p' = 1 - \frac{(1 + e)(a^2 + 3k^2)}{3(a^2 + k^2)} + \frac{(1 + e)^2(a^4 + 2a^2k^2 + 3k^4)}{6(a^2 + k^2)^2}$$

$$+ \frac{(1 + e)^3 a^2 k^4}{3(a^2 + k^2)^2((1 - e)a^2 + 2k^2)}$$

p' is the ratio of the whole energy after collision to the whole energy before collision on an average of all possible cases.

In the same way we get by subtracting v'^2 from u'^2 and putting

$$u'^2 - v'^2 = q'(u^2 - v^2)$$

$$q' = 1 - \frac{(1 + e)(a^2 + 3k^2)}{3(a^2 + k^2)} + \frac{(1 + e)^2(\frac{2}{5}a^4 + 2a^2 k^2 + 3k^4)}{6(a^2 + k^2)^2}$$

$$+ \frac{(1 + e)^3 a^2 k^4}{6(a^2 + k^2)^2(4(a^2 + k^2) - \frac{7}{5}a^2(1 + e))}$$

When $e = 1$ we find $p' = 1$ or the energy is the same before and after impact. When $e = 0$ and $k^2 = \frac{2}{5}a^2$ as in a solid sphere.

$$p' = \frac{1803}{2646} = \frac{601}{882} \qquad \frac{1}{p'} = 1.467$$

$$q' = \frac{1837}{3087} \qquad \frac{1}{q'} = 1.681.$$

22. [Additional Notes on Velocities before and after Impact][a]

Cambridge University Library, Maxwell Manuscripts.

[Start with] equations

$$u_1' = u_1 + \frac{1 + e}{M_1} X \qquad p_1' = p_1 + a_1 \frac{(1 + e)}{M_1 k_1^2}(mZ - nY)$$

We find

$$u_1' = u_1 + \frac{(1 + e)M_2 k_1^2 k_2^2}{M_2 k_2^2(a_1^2 + k_1^2) + M_1 k_1^2(a_2^2 + k_2^2)} \left\{ u_2 - u_1 \right.$$

$$+ \frac{M_2 k_2^2 a_1^2 + M_1 k_1^2 a_2^2}{k_1^2 k_2^2(M_1 + M_2)} l(l(u_2 - u_1) + m(v_2 - v_1) + n(w_2 - w_1))$$

$$\left. + m(a_1 r_1 + a_2 r_2) - n(a_1 q_1 + a_2 q_2) \right\}$$

$$p_1' = p_1 + \frac{(1 + e)M_2 k_2^2 a_1}{M_2 k_2^2(a_1^2 + k_1^2) + M_1 k_1^2(a_2^2 + k_2^2)} \left\{ m(w_2 - w_1) \right.$$

$$- n(v_2 - v_1) - (m^2 + n^2)(a_1 p_1 + a_2 p_2) + lm(a_1 q_1 + a_2 q_2)$$

$$\left. + ln(a_1 r_1 + a_2 r_2) \right\}$$

The equations for the other components of velocity and rotation may be easily written down.

We have next to determine the relations between the energy of agitation before and after impact in the case of the particles of Saturn's rings. The ⟨motions of⟩ energy of agitation is increased between each impact on account of the interpenetration of the particles from different parts of the ring which ⟨move⟩ revolve about Saturn with different velocities. The ⟨motion⟩ energy of rotation of each particle is not influenced by this cause and since we suppose that the motion of agitation is in a steady state we must make the average energy of rotation the same before and after impact.

In order to obtain relations between the energies of agitation in different directions we shall suppose the axes of x, y and z to be the principal axes of agitation, then all products of ⟨the form uv unlike⟩ velocities will disappear and the expressions will depend only on squares of velocities, thus, if we understand that the squares of the quantities refer to the sum of all such squares

$$u_1'^2 = u_1^2 - \frac{2(1+e)M_2 k_1^2 k_2^2}{M_2 k_2^2 (a_1^2 + k_1^2) + M_1 k_1^2 (a_1^2 + k_1^2)}$$

$$\times \left\{ 1 + l^2 \frac{M_2 k_2^2 a_1^2 + M_1 k_1^2 a_2^2}{k_2^2 k_1^2 (M_1 + M_2)} \right\} u_1^2$$

$$+ \frac{(1+e)M_1^2 k_1^4 k_2^4}{(M_2 k_2^2 (a_1^2 + k_1^2) + M_1 k_1^2 (a_2^2 + k_2^2))^2} \left\{ u_2^2 + u_1^2 \right.$$

$$+ \frac{M_2 k_2^2 a_1^2 + M_1 k_1^2 a_2^2}{k_1^2 k_2^2 (M_1 + M_2)} l^2 (u_2^2 + u_1^2)$$

$$+ \left[\frac{M_2 k_2^2 a_1^2 + M_1 k_1^2 a_2^2}{k_1^2 k_2^2 (M_1 + M_2)} \right]^2 (l^4 (u_1^2 + u_2^2) + l^2 m^2 (v_1^2 + v_2^2)$$

$$+ l^2 n^2 (w_1^2 + w_2^2) + m^2 (a_1^2 r_1^2 + a_2^2 r_2^2) + n^2 (a_1^2 q_1^2 + a_2^2 q_2^2)) \right\}$$

$$p_1'^2 = p_1^2 - \frac{2(1+e)M_2 k_2^2 a_1^2}{M_2 k_2^2 (a_1^2 + k_1^2) + M_1 k_1^2 (a_2^2 + k_2^2)} (m^2 + n^2) p_1^2$$

$$+ \left[\frac{(1+e)M_2 k_2^2 a_1^2}{M_2 k_2^2 (a_1^2 + k_1^2) + M_1 k_1^2 (a_2^2 + k_2^2)} \right]^2$$

$$\times \left\{ (m^4 + 2m^2 n^2 + n^4)(a_1^2 p_1^2 + a_2^2 p_2^2) \right.$$

$$+ l^2 m^2 (a_1^2 q_1^2 + a_2^2 q_2^2) + l^2 n^2 (a_1^2 r_1^2 + a_2^2 r_2^2)$$

$$+ m^2 (w_1^2 + w_2^2) + n^2 (v_1^2 + v_2^2) \right\}$$

Now the mean value of l^2, m^2 and n^2 is $1/3$ integrating over the surface of a sphere and the mean value of l^4, m^4, n^4 is $1/5$ and that of $m^2 n^2$, $n^2 l^2$, and $l^2 m^2$ is $1/15$.

a. These notes appear to be an alternative version of document 21.

23. "Mathematical Theory of Saturn's Rings"

Cambridge University Library, Maxwell Manuscripts.

We assume the Rings to consists of independent portions of solid matter of small ⟨size⟩ dimensions compared with the thickness of the rings (100 miles or less) revolving about Saturn in orbits very nearly circular, the differences of the actual motion from that of a satellite in a circular orbit in the plane of the ring being small quantities of the first order. Let S be the mass of Saturn r the distance from his centre then the linear velocity of the supposed satellite at distance r would be $\sqrt{S(1/r)}$ in the positive direction at right angles to r and the actual motion of a particle may be ⟨written⟩ resolved into velocities

u in the direction of r, away from Saturn $= dr/dt$

$\sqrt{S(1/r)} + v$ in the positive direction around Saturn $= r\, d\theta/dt$

w normal to the plane of reference $= dz/dt$

Let z be the actual distance of the particle from the plane of reference ⟨measured perpendicular in the same⟩.

First, for the motion in the plane of reference we have as usual

$$\frac{1}{r} = \frac{S}{h^2}(1 + e\cos(\theta - \alpha)), \qquad \left\langle \frac{1}{r_1} = \frac{S}{h^2}(1 + e\cos\alpha) \right\rangle$$

$$\frac{1}{r^2}\frac{dr}{dt} = \frac{S}{h^2}e\sin(\theta - \alpha)\frac{d\theta}{dt}, \qquad \left\langle \frac{1}{r_1{}^2}\frac{dr_1}{dt} = -\frac{S}{h^2}e\sin\alpha\frac{d\theta}{dt} \right\rangle$$

$$h = r^2\frac{d\theta}{dt}, \qquad \left\langle h = r_0{}^2\frac{d\theta}{dt} \right\rangle$$

Whence

$$u = \frac{S}{h}e\sin(\theta - \alpha)$$

$$v = \frac{1}{2}\frac{S}{h}e\cos(\theta - \alpha)$$

Let u_0 and v_0 be the values of u and v when $\theta = 0$.

$$u = u_0 \cos \theta + 2v_0 \sin \theta$$

$$v = v_0 \cos \theta - \tfrac{1}{2} u_0 \sin \theta$$

In these expressions v is the tangential velocity ⟨referred⟩ relative to that in a circle at the distance from the centre at which the particle is *at that instant* so that while $\sqrt{S/r}$ represents the average tangential velocity v represents the additional velocity due to agitation. ⟨Let us suppose the ring to consist of a system of particles⟩

Let us suppose that R is the total mass of the ring contained within the radius r so that R is a function of r which is zero when r is less than the inner radius of the ring and equal to the whole ring when r is greater than the outer radius of the ring.

Then the quantity of matter in unit of area of the ring will be

$$\frac{1}{2\pi r} \frac{dR}{dr}$$

The attraction of the ring on a particle at its surface, normal to the ring will be

$$\frac{1}{r} \frac{dR}{dr}$$

and if the ⟨matter⟩ particles of the ring be supposed uniformily scattered within a stratum whose thickness $= Z$ on either side of the plane of reference the attraction on a particle within it will be

$$\frac{1}{r} \frac{dR}{dr} \frac{z}{Z}$$

towards the plane of reference. The equation of motion in z will therefore be

$$\frac{d^2 z}{dt^2} = -\left(\frac{S}{r^3} + \frac{1}{rZ} \frac{dR}{dr} \right) z$$

whence if we make

$$1 + \frac{r^2}{SZ} \frac{dR}{dr} = x^2 \mu^2 \text{ and } n = \sqrt{S/r^2} = d\theta/dt$$

we shall have

$$z = z_0 \cos \mu\theta + \frac{w_0}{n\mu} \sin \mu\theta$$

$$w = w_0 \cos \mu\theta + z_0 n\mu \sin \mu\theta$$

z, the distance from the plane of reference and w the velocity perpendicular to that plane are connected together by the following equation which we obtain by eliminating θ from the two equations above

$$z^2 n^2 \mu^2 + w^2 = z_0{}^2 n^2 \mu^2 + w_0{}^2$$

If z^2 represents the mean square of the distance of all the particles from the plane of reference and w^2 the mean square of the agitation normal to it it will be seen that in the case in which the mean thickness of the ring does not alter the value of w^2 will also be constant, that is, for the case of ⟨permanent⟩ stability

$$z^2 = z_0{}^2 \qquad \text{and} \qquad w^2 = w_0{}^2$$

24. "To find an expression for the number of particles which are struck in unit of time and for the proportion of these which describe an angle θ round the central body before being struck again"

Cambridge University Library, Maxwell Manuscripts.

Let x be the number of particles which describe an angle θ without being struck—then while they are describing the additional angle $d\theta$, a number of these will be struck depending on x on the distribution of particles and on $d\theta$ which may be expressed by

$$-dx = \frac{x}{\lambda} d\theta$$

whence

$$x = C e^{-\theta/\lambda}$$

Let $N =$ the whole number of particles then the number struck in unit of time will be

$$\frac{N}{\lambda} \frac{d\theta}{dt}$$

and the number of these which reach a distance θ without being struck will be

$$x = \frac{N}{\lambda} \frac{d\theta}{dt} e^{-\theta/\lambda}$$

and the number of these which will be struck between θ and $\theta + d\theta$ will be

$$-dx = \frac{N}{\lambda^2} \frac{d\theta}{dt} e^{-\theta/\lambda}$$

λ is the mean value of the angle described by a particle between successive collisions.

To find expressions for the $\langle$mean$\rangle$ integrals of $\frac{1}{2}Mu^2$, $\frac{1}{2}Mv^2$, $\frac{1}{2}Mw^2$, $\frac{1}{2}Mz^2$ and Muv in terms of the energies of agitation along the principal axes and the inclination of these axes to the radius vector.

Let a be the sum and b the difference of the energies of agitation in the principal axes in the plane of the orbit and let α be the angle between the radius vector and the greater axis measured in the direction of rotation. Let c be the energy in the normal direction. Then by the ordinary investigation of moments of inertia, internal pressures, etc, etc,

$$\frac{1}{2}\sum Mu^2 = \frac{1}{2}a + \frac{1}{2}b\cos 2\alpha$$

$$\frac{1}{2}\sum Mv^2 = \frac{1}{2}a - \frac{1}{2}b\cos 2\alpha$$

$$\frac{1}{2}\sum Mw^2 = c$$

$$\sum Muv = \frac{1}{2}b\sin 2\alpha$$

25. "To find the relation between the nature of the agitation of the particles just after being struck and just before the next collision"

Cambridge University Library, Maxwell Manuscripts.

The values of u^2, v^2, w^2 and uv are connected by the following equations

$$u^2 = u_0{}^2\cos^2\theta + 4v_0{}^2\sin^2\theta + 4u_0v_0\sin\theta\cos\theta \tag{1}$$

$$v^2 = v_0{}^2\cos^2\theta + \frac{1}{4}u_0{}^2\sin^2\theta - u_0v_0\sin\theta\cos\theta \tag{2}$$

$$w^2 = w_0{}^2\cos^2\mu\theta + n^2\mu^2z_0{}^2\sin^2\mu\theta - 2n\mu z_0w_0\sin\mu\theta\cos\mu\theta \tag{3}$$

$$z^2 = z_0{}^2\cos^2\mu\theta + \frac{1}{n^2\mu^2}w_0{}^2\sin^2\mu\theta + \frac{z}{n\mu}z_0w_0\sin\mu\theta\cos\mu\theta \tag{4}$$

$$uv = (2v_0 - \tfrac{1}{2}u_0)\sin\theta\cos\theta + u_0v_0(\cos^2\theta - \sin^2\theta) \tag{5}$$

$$wz = \left(\frac{w_0{}^2}{n\mu} - z_0{}^2n\mu\right)\sin\mu\theta\cos\mu\theta + z_0w_0(\cos^2\mu\theta - \sin^2\mu\theta) \tag{6}$$

From these we obtain

$$u^2 + 4v^2 = u_0{}^2 + 4v_0{}^2$$

$$w^2 + n^2\mu^2z^2 = w_0{}^2 + n^2\mu^2z_0{}^2$$

which are independent of θ.

In order to get the energy of the agitation of the particles projected in unit of time we must multiply the left side of each equation by

$$\frac{\frac{1}{2}NM}{\lambda}\frac{d\theta}{dt}$$

and the right hand side by

$$\frac{\frac{1}{2}NM}{\lambda^2}\frac{d\theta}{dt}e^{-\theta/\lambda}\,d\theta$$

and integrate from $\theta = 0$ to $\theta = \infty$. It will be sufficient to multiply the right side by

$$\frac{1}{\lambda}e^{-\theta/\lambda}\,d\theta$$

and integrate remembering that

$$\int_0^\infty \sin m\theta\, e^{-\theta/\lambda}\,d\theta = \lambda^2 m/(1 + \lambda^2 m^2)$$

and

$$\int_0^\infty \cos m\theta\, e^{-\theta/\lambda}\,d\theta = \lambda/(1 + \lambda^2 m^2)$$

We thus get from the equations (1), (2) etc the following

1   $(4\lambda^2 + 1)u^2 = (2\lambda^2 + 1)u_0^2 + 8\lambda^2 v_0^2 + 4\lambda u_0 v_0$

2   $(4\lambda^2 + 1)v^2 = (2\lambda^2 + 1)v_0^2 + \frac{1}{2}\lambda^2 u_0^2 - \lambda u_0 v_0$

3   $(4\mu^2\lambda^2 + 1)w^2 = (2\mu^2\lambda^2 + 1)w_0^2 + 2\mu^4\lambda^2 n^2 z_0^2 - 2\mu^2\lambda n w_0 z_0$

4   $(4\mu^2\lambda^2 + 1)z^2 = (2\mu^2\lambda^2 + 1)w_0^2 + 2\dfrac{\lambda^2}{n^2}w_0^2 + 2\dfrac{\lambda}{n}w_0 z_0$

5   $(4\lambda^2 + 1)uv = (2v_0^2 - \frac{1}{2}u_0^2)\lambda + u_0 v_0$

6   $(4\mu^2\lambda^2 + 1)wz = \left(\dfrac{w_0^2}{n\mu} - z_0^2 n\mu\right)\mu\lambda + w_0 z_0$

From (4) and (6) we find that if $z^2 = z_0^2$, $wz = 0$ and since the equations must be true independent of value of λ, $w_0 z_0 = 0$ whence we find by (4) that $w^2 = w_0^2 = n^2\mu^2 z^2$.

26. "To express these relations in terms of a, b, c, and α"

Cambridge University, Maxwell Manuscripts.

Let these quantities be written a_0, b_0, c_0, and α_0, when they refer to the particles when first projected, We have two equations independent of λ

$$5a - 3b\cos 2\alpha = 5a_0 - 3b_0\cos 2\alpha_0$$

$$c = c_0 = n^2\mu^2 z^2$$

(1)–(2) gives

$$(4\lambda^2 + 1)2b\cos 2\alpha = \frac{15}{2}\lambda^2 a_0 + \left(2 - \frac{a}{2}\lambda^2\right)b_0\cos 2\alpha_0 + 5\lambda b_0\sin 2\alpha_0$$

$$(4\lambda^2 + 1)b\sin 2\alpha = \tfrac{3}{2}\lambda a_0 - \tfrac{5}{2}\lambda b_0\cos 2\alpha_0 + b_0\sin 2\alpha_0$$

whence

$$(2\cos 2\alpha - 5\lambda\sin 2\alpha)b = 2b_0\cos 2\alpha_0$$

To find the values of a_0, b_0, c_0, α_0, and λ when the motion of agitation is exactly sustained.

Let us suppose that the result of the collision, on an average of all possible cases is to reduce the total energy of agitation in the ratio of p to 1 and to reduce the *difference* of energy in any two principal axes in the ratio of q to 1, while the direction of these principal axes remains unchanged. We shall calculate the numerical values of p and q by a separate investigation. We then have

$$a + c = p(a_0 + c_0)$$

$$b = qb_0$$

$$\tfrac{1}{2}a - c = q(\tfrac{1}{2}a_0 - c_0)$$

whence

$$a = \tfrac{1}{3}[2p(a_0 + c_0) + q(a_0 - 2c_0)]$$

$$b = qb_0$$

$$c = \tfrac{1}{3}[p(a_0 + c_0) - q(a_0 - 2c_0)]$$

Substituting these values of a, b, c and omitting the suffixes, remembering that a, b, c now ⟨denote the velocities at⟩ refer to the velocities at projection only

$$5[(2p + q - 3)a + 2(p - q)c] = q(q - 1)b \cos 2\alpha$$

$$(p + 2q)c = (3 + q - p)a$$

$$[(4\lambda^2 + 1)q - 1]b \sin 2\alpha + \tfrac{5}{2}\lambda b \cos 2\alpha = \tfrac{3}{2}\lambda a$$

$$2(q - 1)\cos 2\alpha = 5\lambda \sin 2\alpha$$

Whence we find

$$a = 3\lambda(q - 1)(p + 2q)h$$

$$c = 3\lambda(q - 1)(3 + q - p)h$$

$$b \sin 2\alpha = 2(q - 1)[p(3q + 1) - 4q]h$$

$$b \cos 2\alpha = 5\lambda[p(3q + 1) - 4q]h$$

where h is a ⟨constant⟩ quantity not yet determined. We also find the following equations in λ^2, p and q

$$[\lambda^2(16q^2 - 16q + 25) + 4(q - 1)^2][p(3q + 1) - 4q]$$

$$= 9\lambda^2(q - 1)(p + 2q)$$

(Whence $\lambda^2 = 4(q - 1)^2(3q + 1)p - 16q$)) [scored through]

$$\lambda^2 = 4(q - 1)^2 \frac{(3q + 1)p - 4q}{(64q^2 + 18q)(q - 1) + 100q - [(3q + 1)16q(q - 1) + 25(3q + 1) - 9(q - 1)]p}$$

$$= 2(q - 1)^2 \frac{(3q + 1)p - 4q}{32q^3 - 23q^2 + 41q - [24q^3 - 16q^2 + 25q + 17]p}$$

27. [Work Sheets]^a

Cambridge University Library, Maxwell Manuscripts.

$$a = \lambda q(p + 2q - 3)$$
$$c = 2\lambda q(q - p)$$

$$b\cos 2\alpha = 5\lambda q(p - 1)$$
$$b\sin 2\alpha = 2(q - 1)(p - 1)$$

$$25\lambda^2 \qquad +4$$

$$(4\lambda^2 + 1)2p(2a + c) = (2\lambda^2 + 1)\,4a + (4\lambda^2 + 1)2c + 17\lambda^2 a - 15\lambda^2 b\cos 2\alpha$$
$$+ 3\lambda b\sin 2\alpha$$

$$(4\lambda^2 + 1)4p\lambda q(3q - 3) = (25\lambda^2 + 4)\lambda q(p + 2q - 3) + 4\lambda q(4\lambda^2 + 1)(q - p)$$
$$- 75\lambda^3 q(p - 1) + 6\lambda(q - 1)(p - 1)$$

$$\lambda^2\{48pq(q - 1) - 25q(p + 2q - 3) - 16q(q - p) + 75q(p - 1)\}$$
$$+ \{+12pq(q - 1) - 4q(p + 2q - 3) - 4q(q - p) - 6(q - 1)(p - 1)\} = 0$$

4

$$5(a' - a) = 3(b' - b)\cos 2\alpha \qquad\qquad a = (p + 2q - 3)f$$

$$\frac{5}{3}a(2p + q - 3) - \frac{5}{3}c(q - 1) = 3(q - 1)b\cos 2\alpha \qquad\qquad c = 2(q - p)f$$

$$\frac{5}{3}f(9(p - 1)(q - 1)) = 3(q - 1)b\cos 2\alpha \qquad\qquad b\cos 2\alpha = 5(p - 1)f$$

$$5f(p - 1) = b\cos 2\alpha$$

3—

$$((4\lambda^2 + 1)q - 1)b^4\sin 2\alpha = \tfrac{1}{2}\lambda\{3(p + 2q - 3) - 25(p - 1)\}f$$

$$6q - 22p + 16$$

$$\{4\lambda^2 q + (q - 1)\}b^4\sin 2\alpha = \tfrac{1}{2}\lambda(8 + 3q - 11p)f'$$

2(1 − 2)

$$4(4\lambda^2 + 1)b'\cos 2\alpha = 4(2\lambda^2 + 1)b\cos 2\alpha + 15\lambda^2 a - 17\lambda^2 b\cos 2\alpha + 5\lambda b\sin \alpha$$
$$\{4(4\lambda^2 + 1)q + 9\lambda^2 - 4\}5(p - 1)f = 15\lambda^2(p + 2q - 3)f + 5\lambda b\sin \alpha$$
$$\{[\lambda^2(16q + 9) + 4(q - 1)](p - 1) - 3\lambda^2(p + 2q - 3)\}f = \lambda b\sin \alpha$$
$$\lambda^2[16pq - 16q + 9p - 9 - 6q - 3p + 9] + (p - 1)(q + 1) = \lambda b\sin \alpha$$

$$\lambda^2[16pq - 22q + 6p] + 4(p-1)(q-1) = \frac{\lambda b \sin 2\alpha}{f}$$

$$[4\lambda^2 q + (q-1)]b \sin 2\alpha = \lambda(8 + 3q - 11p)f$$

$$8\lambda^4 q\{8pq - 11q + 3p\} + 16\lambda^2 q(p-1)(q-1)$$
$$+\, 2\lambda^2(q-1)(8pq - 11q + 3p) + 4(p-1)(q-1)^2$$
$$-\, \lambda^2(8 + 3q - 11p)$$

$$\lambda^2(8pq - 11q + 3p) + 2(p-1)(q-1) = \frac{\lambda b \sin 2\alpha}{2f}$$

$$\lambda^2(8 + 3q - 11p)$$

$$= (4\lambda^2 q + (q-1))\frac{\lambda b \sin 2\alpha}{f}$$

$$= (4\lambda^2 q - (q - \tfrac{1}{2}))\frac{\lambda b \sin 2\alpha}{2f}$$

$$2(4\lambda^2 + 1)(p-1)(q-1) \qquad\qquad a' = a\frac{2p+q}{3} + c\left(\frac{p-q}{3}\right)$$

$$u^2 = a + b\cos 2\alpha \qquad 2a' + c' = p(2a + c)$$

$$v^2 = a - b\cos 2\alpha \qquad a' - c' = q(a - c) \qquad c' = a\frac{2p - 2q}{3} + c\frac{p + 2q}{3}$$

$$uv = b \sin 2\alpha \qquad\qquad b' = qb \qquad\qquad b' = qb$$
$$z^2 = c$$

$$(4\lambda^2 + 1)(a' + b'\cos 2\alpha) = (2\lambda^2 + 1)(a + b\cos 2\alpha) + 8\lambda^2(a - b\cos 2\alpha) + 4\lambda b \sin 2\alpha$$
$$(4\lambda^2 + 1)(a' - b'\cos 2\alpha) = (2\lambda^2 + 1)(a - b\cos 2\alpha) + \tfrac{1}{2}\lambda^2(a + 2\cos 2\alpha) - \lambda b \sin 2\alpha$$
$$(4\lambda^2 + 1)b' \sin 2\alpha = 3\tfrac{1}{2}\lambda(3a - 5b\cos 2\alpha) + b \sin 2\alpha$$
$$c' = c$$

- - - - - - - - - -

$$a\frac{2}{3}(p-q) + c\frac{p+2q}{3} = c \qquad c = 2(q-p)f \qquad a = (p + 2q \cdot 3)f$$

$$(4\lambda^2 + 1)2qb\cos 2\alpha = (2\lambda^2 + 1)2b\cos 2\alpha + 7\tfrac{1}{2}a\lambda^2 - 8\tfrac{1}{2}b\lambda^2 \cos 2\alpha + 5\lambda b \sin 2\alpha$$

$5\lambda(3)$

$(4\lambda^2+1)5\lambda qb\sin 2\alpha = 7\tfrac{1}{2}a\lambda^2 - 12\tfrac{1}{2}\lambda^2 b\cos 2\alpha + 5\lambda b\sin 2\alpha$

$2(q-1)b\cos 2\alpha = 5\lambda qb\sin 2\alpha \qquad b\cos 2\alpha = 5\lambda qg \qquad b\sin 2\alpha = 2(q-1)g$

$(1+2+4) \qquad (4\lambda^2+1)p(2a+c) = (2\lambda^2+1)2a + (4\lambda^2+1)c + 8\tfrac{1}{2}\lambda^2 a - 7\tfrac{1}{2}\lambda^2 b\cos 2\alpha + 3\lambda b\sin 2\alpha$

$a\{2p(4\lambda^2+1) - (12\tfrac{1}{2}\lambda^2+2)\} - (4\lambda^2+1)c = 3\lambda b\sin 2\alpha - 7\tfrac{1}{2}\lambda^2 b\cos 2\alpha$

$1+4.2 \qquad \{(p+2q-3)(2\lambda^2(8p-12\tfrac{1}{2})+2(p-1)) - (4\lambda^2+1)(2q-2p)\}f = \{6\lambda(q-1) - 12\tfrac{1}{2}\lambda^3 q\}g$

$(4\lambda^2+1)(5a'-3b'\cos 2\alpha) = (2\lambda^2+1)(5a-3b\cos 2\alpha) + \lambda^2(10a-6b\cos 2\alpha)$

$5a'-3b'\cos 2\alpha = 5a-3b\cos 2\alpha$

$\tfrac{5}{3}a(2p+q) - \tfrac{5}{3}c(q-p) - 3qb\cos 2\alpha = 5a-3b\cos 2\alpha$

$\tfrac{5}{3}\{(2p+q-3)(p+2q-3) - 2(q-p)^2\}f = 3(q-1)5\lambda qg$

$q(pq-p-q+1)$

$a = (p+2q-3)\lambda q$

$c = 2(q-p)\lambda q \qquad \lambda^3 q$

$b\cos 2\alpha = 5(p-1)\lambda q$

$b\sin 2\alpha = 2(q-1)(p-1)$

$(p-1)f = \lambda qg \qquad f = \lambda qh \qquad g = (p-1)h$

$$8p^2 - 12\tfrac{1}{2}p + 16pq - 25q + 37\tfrac{1}{2} + 2p^2 q + 4pq^2 - 3pq$$

$$\begin{array}{ll} -24p & -8q \\ +8p & \\ +12\tfrac{1}{2}p & \end{array} \quad \begin{array}{l} -2pq - 4q^2 + 6q \\ -2pq - 2q^2 \\ +2pq - 2q^2 \\ -6pq + 6p + 6q - 6 \end{array} \quad \left.\right) \; \lambda\langle q\rangle$$

$-12\tfrac{1}{2}$

$$8p^2 - 16p + 16pq - 33q + 25 \qquad 2p^2 q + 6pq^2 - 9pq - 6q^2 + 12q + 6p - 6$$

$(4\lambda^2+1)q^2(q-1)(p-1) = \tfrac{1}{2}\lambda^2 q(3p+6q-9-25p+25) + 2(q-1)(p-1)$

$(4\lambda^2+1)4q(q-1)^2(p-1) = \lambda^2 q(6q+16-22p) + 4(q-1)(p-1)$

$\lambda^2 q(16pq - 16p - 16q + 16 + 4(q-1)(p-1)) = 0$

$+ 22p - 6q - 16$

$\lambda^2 q(16pq + 6p - 22q) + 4(q-1)(p-1) = 0$

a. These sheets are obviously where Maxwell worked out the algebraic steps involved in the equations in the preceding notes and rough draft. Other sheets of more algebraic details exist but the above give a good example of the way Maxwell separated the arithmetical details from the drafts of papers he was preparing.

28. ["Let a particle describe an orbit ..."]

Cambridge University Library, Maxwell Manuscripts.

Let a particle describe an orbit about a centre of force μ/r^2 and let its velocity at projection be $v \cos \phi$ in the radial and $v \sin \phi + V$ in the ⟨radial⟩ tangential direction, where V is the velocity in a circle and v is small compared to V.

Required the radius vector, and the velocity relative to that in a circle after describing an angle θ. Let u, v, and ϕ be the values at projection, u', v', ϕ' afterwards then

$$u' = \frac{\mu}{h^2}(1 + e \cos \alpha \cos \theta + e \sin \alpha \sin \theta)$$

$$\frac{du'}{d\theta} = \frac{\mu}{h^2}(e \sin \alpha \cos \theta - e \cos \alpha \sin \theta) = -\left(\frac{1}{h}\right) v' \cos \phi'$$

$$h = \sqrt{\frac{\mu}{u}\left(1 + \frac{v \sin \phi}{\sqrt{\mu u}}\right)} = \sqrt{\frac{\mu}{u'}\left(1 + \frac{v' \sin \phi'}{\sqrt{\mu u'}}\right)}$$

At projection

$$\theta = 0 \qquad u = \frac{\mu}{h^2}(1 + e \cos \alpha),$$

$$\frac{du}{d\theta} = \frac{\mu}{h^2} e \sin \alpha = -\frac{1}{h} v' \cos \phi'$$

$$e \cos \alpha = \frac{v}{\sqrt{\mu u}}(2 \sin \phi)$$

$$e \sin \alpha = -\frac{v}{\sqrt{\mu u}} \cos \phi$$

$$u' = u\left\{1 - \frac{v}{\sqrt{\mu u}} 2 \sin \phi\right\}\left\{1 + \frac{v}{\sqrt{\mu u}}(2 \sin \phi \cos \theta - \cos \phi \sin \theta)\right\}$$

$$u' = u\left\{1 - \frac{v}{\sqrt{\mu u}}\{2 \sin \phi(1 - \cos \theta) + \cos \phi \sin \theta\}\right\}^{\text{a}}$$

$$v' \cos \phi' = -\sqrt{\mu u}\left[1 - \frac{v}{\sqrt{\mu u}} \sin \phi\right]$$

$$\left[-\frac{v}{\sqrt{\mu u}} \cos \phi \cos \theta - \frac{v}{\sqrt{\mu u}} 2 \sin \phi \sin \theta\right]$$

* $v' \cos \phi' = v(\cos \phi \cos \theta + 2 \sin \phi \sin \theta)^{\text{b}}$

     $v' \sin \phi' = hu' - \sqrt{\mu u'}$

$$= \sqrt{\mu u} \left[\left(1 + \frac{v}{\sqrt{\mu u}} \sin \phi \right) \right.$$

$$\times \left(1 - \frac{v}{\sqrt{\mu u}} (2 \sin(1 - \cos \theta) + \cos \phi \sin \theta) \right)$$

$$\left. - \left(1 - \frac{\frac{1}{2}v}{\sqrt{\mu}} [2 \sin \phi (1 - \cos \theta) + \cos \phi \sin \theta] \right) \right]$$

* $v' \sin \phi' = v(\sin \phi \cos \theta - \frac{1}{2} \cos \phi \sin \theta)$

* $z' = z \cos \theta + \dfrac{dz}{d\theta} \sin \theta$

* $\dfrac{dz'}{d\theta} = \dfrac{dz}{d\theta} \cos \theta - z \sin \theta$

a. This equation in u' does not follow algebraically from the previous expression. It should be

$$u' = u \left\{ 1 - \frac{v}{\sqrt{\mu u}} 2 \sin \phi (1 - \cos \theta) - \frac{v}{\sqrt{\mu u}} 2 \sin \phi \frac{v}{\sqrt{\mu u}} \right.$$

$$\left. \times (2 \sin \phi \cos \theta - \cos \phi \sin \theta) \right\}$$

The first expression for u' was found by substituting expressions for $e \cos \alpha$ and $e \sin \alpha$, then approximating the expression

$$\frac{u}{\left(1 + \dfrac{2v \sin \phi}{\sqrt{\mu u}} \right)}$$

by

$$u \left(1 - \frac{v}{\sqrt{\mu u}} 2 \sin \phi \right).$$

Subsequent expressions that depend on this one for u' are also incorrect.

b. This does not follow algebraically from the previous expression for $v' \cos \phi'$.

Index